SpringerBriefs in Environmental Science

W0081967

More information about this series at http://www.springer.com/series/8868

Čedo Maksimović · Mathew Kurian
Reza Ardakanian

Rethinking Infrastructure Design for Multi-Use Water Services

Čedo Maksimović
Department of Civil and Environmental
  Engineering
Imperial College London
London
UK

Mathew Kurian
Reza Ardakanian
UNU-FLORES
United Nations University
Dresden
Germany

ISSN 2191-5547 ISSN 2191-5555 (electronic)
ISBN 978-3-319-06274-7 ISBN 978-3-319-06275-4 (eBook)
DOI 10.1007/978-3-319-06275-4

Library of Congress Control Number: 2014946371

Springer Cham Heidelberg New York Dordrecht London

Printed on acid-free paper

Springer is part of Springer Science+Business Media (www.springer.com)

Contents

Figures

Tables

Chapter 1
Rethinking Infrastructure Design

Abstract For the first time in history, more than half the world's population will be living in urban areas. Cities offer enormous opportunities, but they also create problems that degrade the environment, thus the quality of life in the city, surrounding suburban areas and downstream settlements. By 2030, it is projected that 4.9 billion people will be living in cities. Managing it in a sustainable way provides a unique opportunity. The objectives of water management in urban areas are to ensure that no damage is caused during extreme participation and that long periods of droughts do not cause problems in cities or the countryside. The Blue Green Dream project promotes a new paradigm for efficient planning and management of the urban environment. This volume introduces the methods by which mutual interactions of urban water infrastructure (blue assets) and urban vegetated areas (green assets) are taken into account in the synergy of spatial planning and optimised modelling of ecosystems' performance indicators. This method of planning should make future developments cheaper to build, their users will pay lower utility bills (for water, energy, heating), such developments will be more pleasant to live in and property value would likely be higher.

Keyword Multiple-Use Water Services · Blue services · Climate change · Green services · Millennium Development Goals · Pluvial flooding · SUDS · UHI · WSUD

1 Introduction

We are reaching the milestone: for the first time in history, more than half the world`s population will be living in urban areas. Cities offer enormous opportunities, but they also create problems that degrade the environment thus the quality of life in the city, surrounding suburban areas and downstream settlements. Humanity is rapidly urbanising. While globally only 220 million people (13 %) lived in urban areas in 1900, this increased to 3.2 billion (49 %) by 2005 and is projected to reach 4.9 billion (60 %) by 2030. If we consider those numbers, we will have major increases in urbanisation as cities become settlements for a majority

© The Author(s) 2015   1
Č. Maksimović et al., *Rethinking Infrastructure Design for Multi-Use
Water Services*, SpringerBriefs in Environmental Science,
DOI 10.1007/978-3-319-06275-4_1

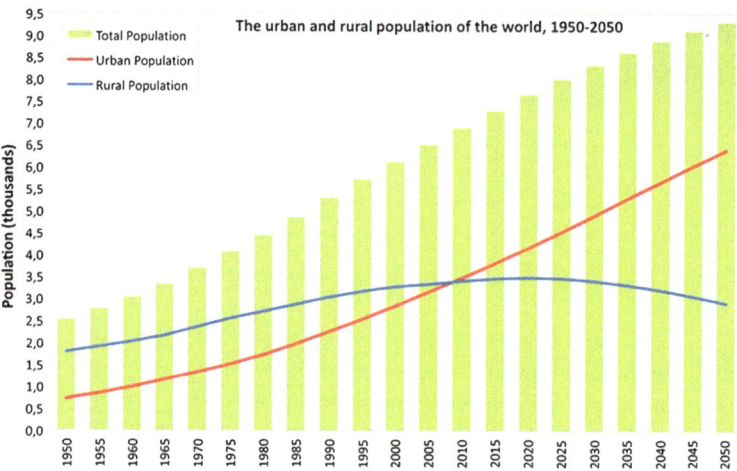

Fig. 1 2009–2010 revision of world urbanisation prospects © (*Source* UN, Population Division of the Department of Economic and Social Affairs)

of the world's population. Thus, growth of urban areas or creation of new ones is inevitable. Some countries with huge populations (India for example) are yet to see this rapid urbanisation happen. If it happens spontaneously (in an unmanaged fashion), huge environmental problems will surface. Managing it in a sustainable way provides a unique opportunity (Fig. 1).

The objectives of water management in urban areas are to ensure that no damage is caused during the extreme participation and that long periods of droughts do not cause problems in cities or the countryside. Our urban centres rely on water and aquatic ecosystems for services, such as oxygen production, carbon storage and natural filtering of toxins and pollutants. Besides the clean water supply for our daily needs, we depend on water to grow our food and produce resources, and to transport our goods and waste. By 2030, 47 % of the world's population will be living in areas of high water stress (Organisation for Economic Co-operation and Development (OECD) 2008). Freshwater ecosystems are among the most degraded on the planet (United Nations Environmental Program (UNEP) 2009). Water shortages, flooding and watercourse pollution are all signs of stress where developed areas have troubled interaction with the natural water cycle and where water has become a risk rather than an opportunity. Urban planners face a choice in their future approach to incorporate appropriately the interactions of water resources and green/vegetated spaces into an innovative concept of future cities. Their cities can become increasingly dependent on rural support areas and enlarge their urban 'shadow', potentially damaging food production, nutrient flows and water resources; or they can shift from being resource users to resource managers (Bahri 2012), altering their consumption patterns, waste management and planning to balance resource flows better to and from cities. Among numerous options, the approach pursued in this tutorial section based on an innovative concept of the Blue Green Dream project.

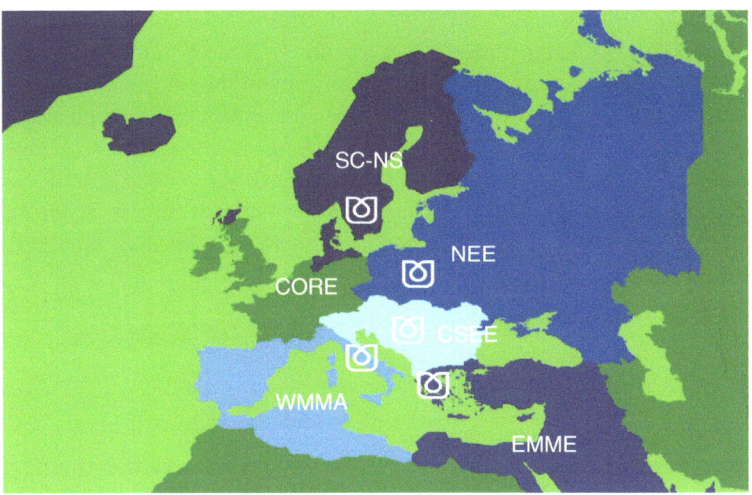

Fig. 2 Five regional BGD centres in Europe

Multiple-use water services (MUS) promotes a new paradigm for efficient planning and management of the urban environment: one that maximises ecosystem services, minimises environmental footprint and increases cities' adaptive capacity to changing climate, demographic and socio-economic conditions. The paradigm aims at enhancing the synergy of urban blue (water) and green (vegetation, energy efficiency) systems and provides effective, multifunctional MUS to support urban adaptation to future climatic changes. Contrary to most energy efficiency, or Green Infrastructure (GI) programmes that are propelled by government, i.e. apply the top-down approach, the MUS is planned to combine both top-down and bottom-up initiative propelled by masses. This approach guarantees resource management efficiency and goal appraisal—of which both are needed for a successful fulfilment of the project. The MUS paradigm is planned to become global and it will be pursued in Europe through the core team (DE, FR, NL and UK) and the network of five regional centres (RC) (Fig. 2). They are under development for the following five regions: Central and South-Eastern Europe (CSEEE), Scandinavia and North Sea, North-Eastern Europe (NEE), Western Mediterranean and Maghreb (WMMA) and Eastern Mediterranean and Middle East (EMME), followed by creation of the global MUS network in 2014. Each of the Regional BGD centres will create a polycentric network of focal points (FP) in each of the countries. The concept of the RC's functioning is presented in Maksimović et al. (2013).

The objectives of MUS are to achieve optimised solutions for urban developments and retrofitting primarily by optimising performance of ecosystems affected by water and greenery interactions. In the case of water, they are based on the concept of **3R**s—reduce, reuse and recycle. While 'reduce' is bounded to behaviour change of users and cannot be integrated within a water management system, 'reuse' and 'recycle' are most certainly topical. The opportunities for improving this urban imbalance lie, among others, at the neighbourhood level. Applying a more natural approach to water management, for instance with thorough use of natural outflows or buffering, more use of on-site infiltration into the soil, better utilisation of rainwater reuse of wastewater in due time and improved urban planning, which incorporates local characteristics into the management plan will be more cost-effective and efficient in comparison to centralised water management. Since the centralised systems are based on artificial balance of urban water, they are becoming increasingly unbalanced and vulnerable to increasing urbanisation and climate change.

2 Climate Change

The most common weather (climate) variables impacting the urban environment are air temperature, precipitation (rainfall) and water level in the seas and rivers. Being stochastic in nature, they are characterised by trends in their mean value and extremes represented by standard deviation. As shown in Fig. 3, there is no agreement in the long-term changes of mean values of air temperature predicting that it may rise between 0.5 and 4 °C. However as experienced all over the world, extremes of both air temperature and precipitation are increasing in both magnitude and durations, causing inter alia more severe droughts and all sorts of floods (pluvial, fluvial, coastal and groundwater). Multiple water usage aims at reducing **vulnerability of urban** areas to these long-term climate changes and variability of extremes.

Fig. 3 Predicted global surface warming (°C) with variability of extremes of rainfall and air temperature superimposed

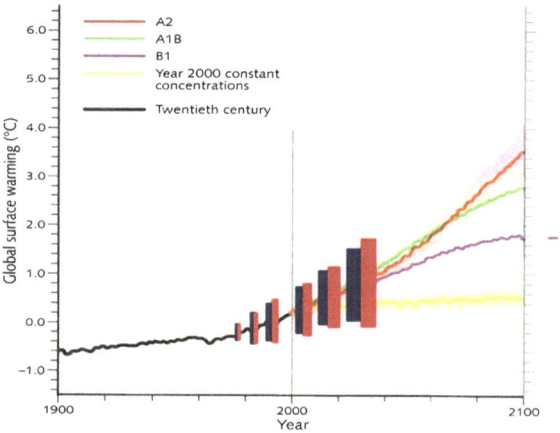

3 What Is Wrong with Existing Systems?

When the problems of climate change became obvious (through extremes), many people thought about our relationship with the environment. Even in the 1970s, questions of population, food production, industrial production, pollution and consumption of non-renewable natural resources rose as problems of non-long-term viability of modern cities were introduced.

Traditional engineering design of structures and systems exposed to impacts of stochastic conditions (variables) was to adapt the acceptable risk (thus costs) to selected return periods. However, the climate changes created issues with this methodology since the systems designed to sustain certain return period are now exposed to the same extreme conditions more frequently. Recently, Fratini et al. (2012) 'customised' this method by introducing so-called Three Points Approach (3PA) to be used in analysis of adaptations to climate changes. If one single function (flood for example) of a technical system were analysed in isolation, in order to adapt for climate changes, one would have to invest heavily into increasing its capacity ('rich men's solution').

Traditional water management, which treats elements of the urban water system as isolated services, has led to an unbalanced urban 'metabolism' (Novotny et al. 2010). It is an inefficient, unsecure and separated centralised system. On one side of the system there is 'production of drinking water supply' (energy and chemically intensive), whose capacities are being exceeded by rapid urbanisation. In extreme cases (California for example), even if almost all water resources from the broader region are captured and used, the system has reached the maximum natural capacity to support it. At the other end of the system is the wastewater treatment plant. Plants that are environmentally unfriendly (impact of sludge and chemically intensive) and use lots of space (from nature and agriculture) are having problems accepting and treating a combined quantity of storm and wastewater. Furthermore, the extent of paved areas inside urban areas, during extreme rain events, generate large amount of polluted run-off that wastewater treatment plants cannot obtain. Thus, untreated water (blend of raw sewerage and storm run-off) is released to surface waters (rivers, ponds, lakes, etc.), which leads to environmental degradation. This process is illustrated in Fig. 4, which depicts the conventional urban drainage systems with and without wastewater treatment plants.

To sum up, the quantity of water in cities is a problem today—both shortage and excess. A new approach is clearly needed. An approach that can resolve water problems, provide multiple environmental benefits and support sustainable development (US Environmental Protection Agency (US-EPA) (2012)). One solution uses natural processes and location characteristics in order to reduce negative footprints cities have on the environment. Integrated urban water management is no one-size-fits-all model nor is one model sufficient. Rather, it reframes a city's relationships to water and other resources, and conceptualises the ways in which they can be overseen. The goals of urban water management are to ensure access to water and sanitation infrastructure and services (in the least developed countries this

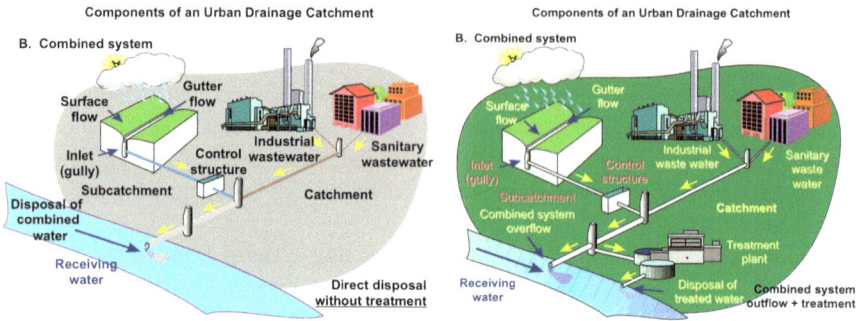

Fig. 4 Conventional urban drainage systems with and without wastewater treatment plants

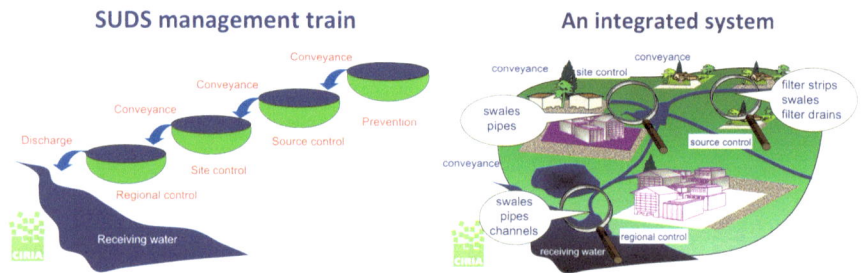

Fig. 5 SUDS management train as part of an integrated storm drainage system

means meeting the UN Millennium Development Goals); manage rainwater, wastewater, storm water drainage and run-off pollution; control waterborne diseases and epidemics; and reduce the risk of water-related hazards, including floods, droughts and landslides. In the innovative concept pursued by the MUS paradigm, this also means activation of new potentials for natural and processed water resources and its by-product (recyclates) in interaction with other urban services. In urban water systems that mean visible water systems, increased water awareness and contributions to water usage reduction; reuse of wastewater and use of rainwater, reduced stress on water resources; reduction of pavement and increased infiltration, increased evapotranspiration, reduced run-off and improved urban area amenities. In this respect important are two principles introduced during the past two decades: (a) SUDS (Sustainable Urban Drainage System) introduced by the UK CIRIA (Construction Industry Research and Information Association—CIRIA 555 (1998), also called BMP in the USA). In its original version, SUDS aimed at reducing surface run-off peaks and volumes, improving water quality and creating urban amenities. The initial version of SUDS is based on the so-called management train (Fig. 5). While broadly implemented in other countries (Sweden, Germany), and although introduced almost 20 years ago, it is only recently being gradually accepted in the UK where it is the right time to be superseded by more

Fig. 6 From natural to developed (built) and back to natural principles in surface run-off

comprehensive concepts; and (b) WSUD (Water-Sensitive Urban Design), initially introduced in Australia (2009) as an extension of SUDS principle, mainly by using rain gardens and biofilters to enhance run-off quality so that it can be safely released into receiving water bodies. WSUD has been successfully implemented in many small-scale projects in Australia, which made many positive changes, particularly in Melbourne. The WSUD concept has been recently introduced in the UK with some new features added to the initial concept of WSUD.

While embracing both SUDS and WSUD, a new concept, with a much broader framework, termed henceforth MUS is proposed within the MUS paradigm as a transition from a 'rich man's solution' to 'a wise man's solution/culture' (Fig. 6).

4 What Is the Multiple-Use Water Services Concept?

MUS Solutions enhance the synergy of urban blue (urban water infrastructure) and urban vegetated areas (green assets) by providing multiple benefits to urban eco-system services and functions. They encompass the options and interventions developed within the WSUD (Water-Sensitive Urban Design) concept but go beyond them (Fig. 7).

The key aspects of urban design are to create a place for people, enrich the existing places and make connections between urban areas. At the same time, designs have to be economically viable and climate (environmental) adaptive. By combining well-established principles of urban planning[1] with a growing under-standing of new technologies and natural principles, urban planners are able to create better and more secure places for future generations. The key issues creating the need for MUS solutions are presented in Fig. 8.

Ecosystem services provide effective, multifunctional support to urban adaptation to future climate changes and variation of climate extremes. During the planning or designing process, architects/planners have the responsibility to take into account the following:

---

[1] Principles such as mix of use and users, plenty of public space, combination of wide streets and mews (inner streets) provide parking spaces that don't dominate street view, etc.

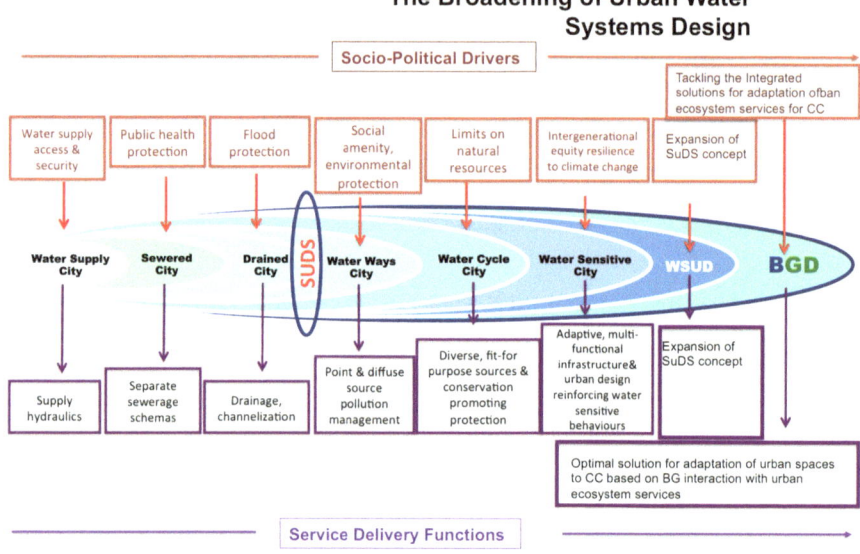

Fig. 7 Development of water management in the cities, adapted from Brown et al. (2008)

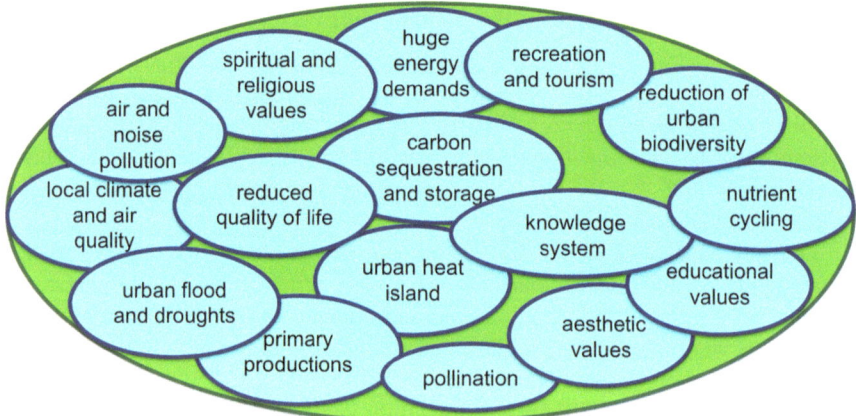

Fig. 8 Issues creating need for integrated MUS solution

- Near zero carbon—reduction of energy use
- Near zero waste
- Sustainable transport
- Local and sustainable materials
- Local and sustainable food
- Sustainable water

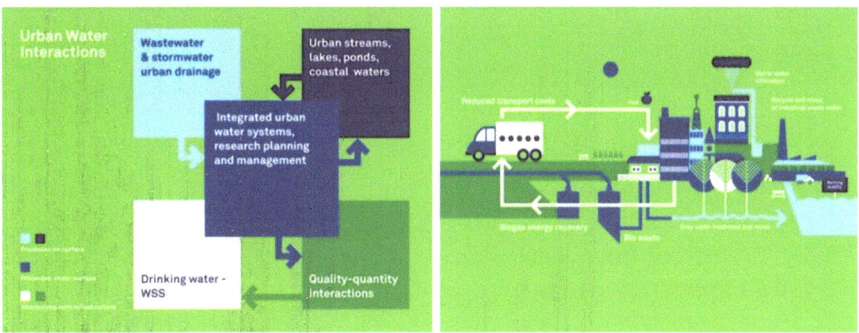

Fig. 9 Illustration of MUS interactions, starting with interactions in the water sector towards interactions with vegetated areas and other ecosystem services, energy, food, etc.

- Natural habitats and wildlife
- Culture and heritage
- Equity and fair trade
- Health and happiness (Fig. 9)

In order to understand better all benefits and interactions provided by solutions and paradigm, solutions will be briefly explained through the following ecosystem services:

- Pluvial flood risk
- Water pollution
- Alternative water sources
- Urban Heat Island (UHI)
- Air and noise pollution
- Urban agriculture
- Urban health, amenities and social behaviour

Here are some of the MUS solutions that have impacted urban areas when the water questions are in place:

- Private green gardens
- Green roofs (both intensive and extensive)
- Porous pavements
- Urban farms/urban agriculture
- Green facades
- Infiltration boxes
- Water roofs
- Cooling with water elements (e.g. fountains and ponds)
- Adding green in streetscape (grass, herbs)
- Rainwater harvesting
- Parks and urban forests
- Water squares

- Artificial urban wetlands
- Deep ground infiltration
- Bio-swales and swales

4.1 Reduced Pluvial Flood Risk

While conventional storm water management relies on a system of dykes (flood defence structures) and network of sewer systems and canals to channel water quickly into rivers or the sea to prevent flooding, MUS is based on the idea that it is better to address the problem where it occurs (at the 'source') (see Fig. 10).

When pluvial flooding is of concern, the biggest issue for cities is massive impermeable surfaces (paved areas, roads, roofs, etc.). Instead of infiltrating, used or evaporating water goes directly into the drainage system. In order to deal with this problem, nature friendly solutions such as green roofs, green facades, rain gardens, bio-swales, retention/detention ponds, wetlands and others are proposed. There is great potential in connecting greenery, water features and the building structures itself to channel water to various drainage features.

Rainwater can be (re)used, water can be harvested from roofs and other sealed surfaces. Although there is strict control on harvesting rainwater for drinking, it is a straightforward procedure to harvest rainwater for irrigation, toilet flushing, washing clothes or for the washing of pavements and walkways. In this way, total run-off from the site is reduced. If combined with the open water bodies, all rainwater can be directed in that way so the problem becomes a useful resource.

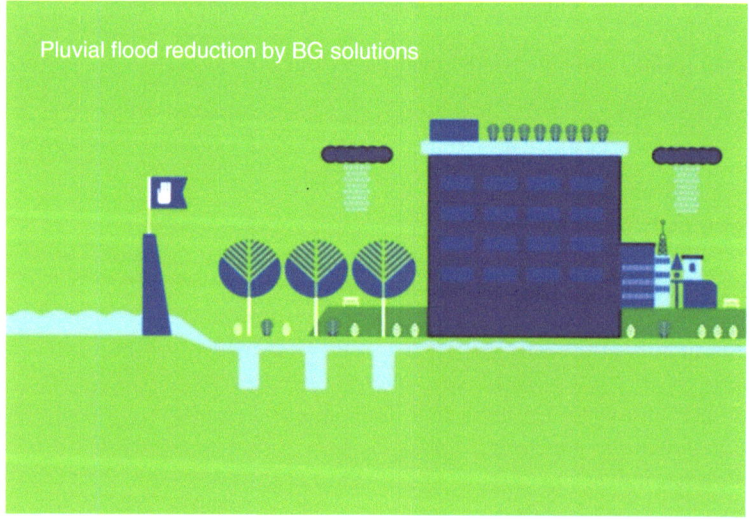

Fig. 10 Pluvial flood risk reduction

Fig. 11 Examples of rain gardens and multifunctional trees (http://www.phillywatersheds.org)

Water storage, both natural (open water bodies such as retention ponds or wetlands) or engineered ones (water tanks or water gardens), serve as run-off control, but also provides opportunity for water demands to be met during longer and more frequent dry periods.

The MUS paradigm is aimed at behaviour and perspective change. People have to realise that water 'drainage' can be something beautiful, watering attractive plants within low maintenance arrangements while providing habitat for wild life and increasing amenity of space (see Fig. 11). In the US, the simple but powerful idea of the multifunctional rain garden is spreading rapidly.

4.2 Water Pollution

What about the quality of water in our water streams? Indeed, most of the urban run-off comes from roofs, streets and paved areas. Pollutants from vehicles including oil, oil combustion products and heavy metals accumulate on sealed surfaces and are washed into drains and ultimately into water streams following rainwater. The water has to be filtrated before discharging. The solutions may be in redirecting run-off water through the soil and vegetation (including tree pits), which would enable natural microbes to remove pollutants before they reach drains or streams. If geology allows, water can be infiltrated to recharge groundwater or overflow to conventional drains, which will reduce stress on sewers and water treatment plants (Fig. 12).

It is a common misconception that sustainable drainage systems may not be possible to deal with the amount of water or that at the least, need additional overflow protection or connection to a drainage system. However, integration of MUS solutions such as rainwater harvesting, green roofs, rain gardens and other features may be enough to retain complete participation run-off. Good example of such solution is Potsdamer Platz in Berlin. The whole area is designed in a way for all run-off to be intercepted and stored in underground storage or waterpark. In 10 years, there has been no run-off from the site.

Fig. 12 One of the possible
uses of harvested roof water

Vegetation purifies water. Nitrates and phosphates can only be absorbed effectively and efficiently through vegetation, and by protracted microbial and bacterial breakdown (Brosens 2008). Green areas in cities reduce the costs of drainage and retention facilities elsewhere and ease the need for water treatment.

4.3 Alternative Water (Re)Sources

Potable water from clean sources is rare, other sources of water must be treated at high cost and the volume of wastewater is growing. Cities produce large quantities of wastewater and other forms of waste. Where waste treatment is inadequate—or, indeed, entirely absent—waste disposal sets in motion a cascade of events that reverberates across a range of ecosystems. Wastewater presents a huge water resource. Different kinds of water can be used for different purposes: freshwater sources (surface water, groundwater, rainwater) and wastewater (black, brown, grey) can be treated to satisfy demands of agriculture, industry and the environment. Water recycling and reuse closes the loop between water supply and wastewater disposal.

In the past, this sounded like an unrealistic illusion. In the future, it will be an essential need and practised as a routine (Fig. 13).

When speaking about water recycling, MUS has two main goals: reduction of potable water usage and the stress on treatment plants. More than half of potable water used in typical household is used for bathing and washing after which it becomes grey water. Grey water can be easily treated on site, which would reduce stresses on water sources. On the other hand, global urbanisation and increased densities in urban areas put a lot of pressure on water treatment plants. The centralised solutions have been questioned. MUS proposed a decentralised approach to water treatment recycling. However, it is not a single one-niche solution. Water

Fig. 13 Examples of harvested roof water and grey water recycling

issue was dealt with by interacting with other ecosystem services. By solving one problem, MUS solutions are designed to provide co-benefits. In the case of purification of water, vegetation plays an important role in the treatment process. The advantage of this decentralised biological treatment system is that they do not produce sludge or require chemicals. In addition, the level of energy consumption is much lower and the operating and management cost are lower as well. Showcases of such systems have been developed in Amsterdam for separate sewer system, which showed not only cheaper but also cleaner effluent.

There are companies that have started producing and marketing simple grey water recycling products.

4.4 Urban Heat Island

As urban areas developed, the landscape changed greatly. As common constructed materials replaced natural ones, more sun energy is absorbed in urban areas. Impermeable and dry areas instead of permeable and moist ones with urban street canyon effect are one of the reasons to contribute to the positive thermal balance of urban areas—Urban Heat Island phenomenon. In climate areas where the population has not developed resilience to heat waves, large-scale fatalities can occur (Fig. 14).

What are the differences between a vegetated scenario and non-vegetated scenario? Two main reasons heat islands are formed are: impermeable and watertight materials, widely used for urban building construction, which directly leads to insufficient moisture available to dissipate the sun's heat; dark materials and urban canyon effect of buildings and pavement 'block the radiation from surfaces to cool sky' (Fig. 15).

Trees and vegetation besides shading can lower the surface temperature through shade and evapotranspiration (evaporative cooling). As in rural areas, trees and vegetation dominate the landscape; urban areas are commonly covered by dry,

Fig. 14 Urban heat island (UHI) mitigation by irrigated vegetation

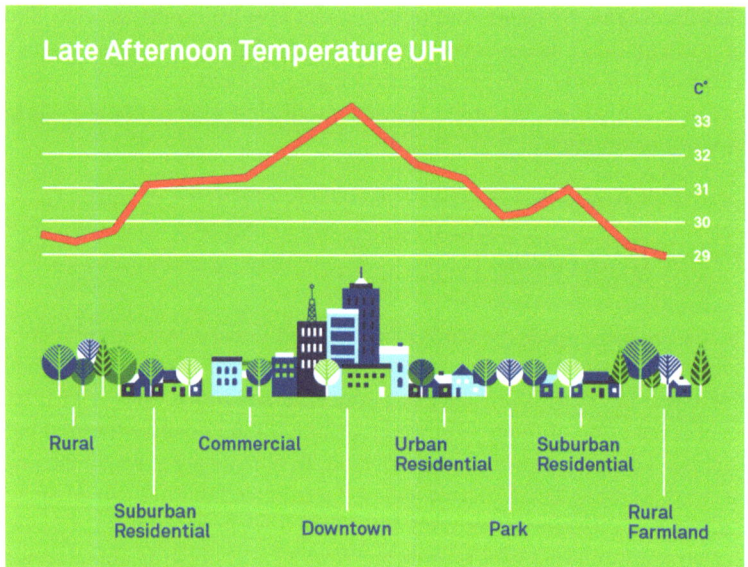

Fig. 15 Sketch of an urban heat island profile (Rosenthal et al. 2008)

impervious surfaces, which lead to less shade and moisture. Therefore, less water is evaporated and surface and air temperature is increased as illustrated in Fig. 16.

Many studies have been carried out to develop the technologies to mitigate the impact of the heat island. Cities have to decrease thermal losses (thermal conductivity) and to control thermal gains (solar radiation). In order to achieve that,

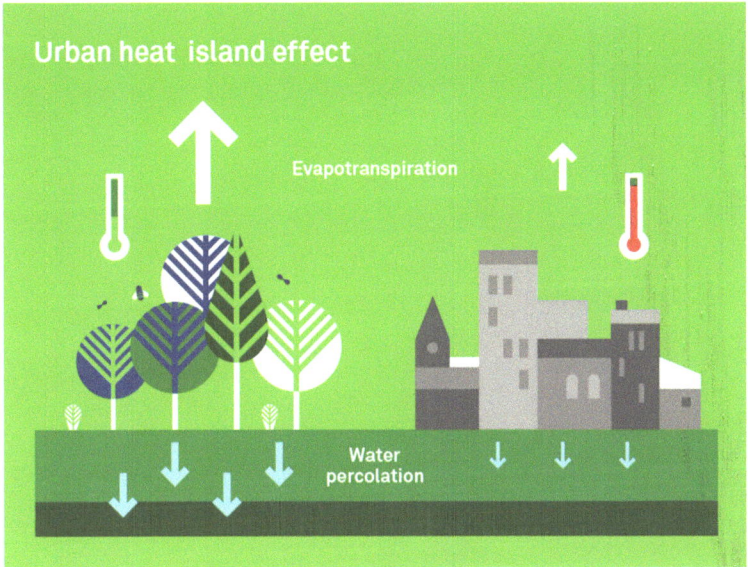

Fig. 16 A simple illustration of the UHI causes

increasing the albedo of the urban environment, increasing urban vegetation coverage and using natural heat sinks are preferred options. Shading and evapotranspiration are the two main techniques that trees and vegetation employ to reduce the temperature. Trees and vegetation are beneficial to the communities in many aspects, including improving thermal comfort, reducing energy usage in buildings, carbon dioxide reduction, improved air quality and decreased storm water run-off.

A Demonstration

A simple demonstration of vegetation influence on the UHI effect is shown on the MSc project conducted on Imperial College South Kensington Campus (Luan 2013). First figure shows the existing state of Imperial College Campus. By greening 100 % of roofs and introducing new tree lines along the sidewalks, the basic model calculated by implementation of green roofs and ground-level vegetation, air temperature is reduced by 2 °C. Achieved temperature reduction can effectively reduce the risk of possible heat-related illnesses and disorder to a lower level, when the air temperature is above 30 °C. However, in order to provide these effects, vegetation has to have constant supply of water so that water is relisted via evapotranspiration. This water does not have to come from potable system. It can come from the BG solution.

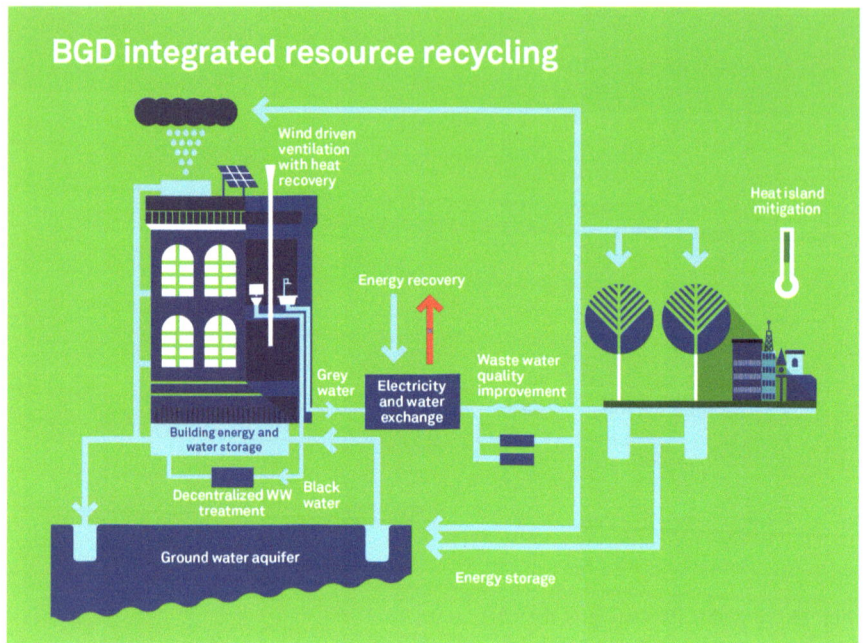

Fig. 17 Combined measures for energy efficiency and UHI mitigation

Results are promising from a pilot project in three US cities (Chicago, Portland and Philadelphia). According to US-EPA (2012), by implementing a green roof at City Hall in Chicago, they have gained positive improvements in: (a) energy saving (in electricity 9,271 and 7,372 KWh/year in terms of heating); (b) improvement in air quality; (c) up to 70 % reduction in storm water load; (d) increase in biodiversity, etc.

The raised temperature in urban areas directly increases energy required for air conditioning and refrigeration in summer. This problem can be mitigated (Fig. 17) by combination of active and passive measures. These may include storing surplus thermal energy in summer (to be recovered in winter together with heat from wastewater) and using the shade and evaporative cooling of (blue) green roofs and walls and tree lines planted in an optimised fashion. Some of these effects can be illustrated by the results of monitoring in Florida to measure the energy saving from urban tree planting. It was found that up to 50 % of cooling-electricity is saved for a building after the addition of trees and shrubs (Meerow and Black 2005). By reducing the need for air conditioning, anthropogenic (manmade) heat is also reduced. That means that, as a feedback, less heat is pumped back into the urban environment.

By combining MUS solutions in the streets and on buildings the properties of urban materials such as albedo, thermal regime, emissivity and specific heat capacity can be drastically improved. The ones that influence solar radiation; emissivity and absorption, thus reduce the UHI effect.

Fig. 18   Picture of a city no tourist would like to take home

4.5 *Air Pollution*

The air quality in urban areas is in constant degradation in many cities (Fig. 18), much of which is caused by human activities: vehicle and industrial emissions. Air pollution leads to health problems, degradation of nature and human environment, has a negative influence on UHI, causes material degradation and of course reduces human comfort.

Just as the vegetation on the ground level does, plants on green roofs can remove several types of air pollutants from the air, such as particulate matter, nitrogenous compounds and sulphur oxides. And due to the photosynthesis of plants, oxygen is produced and carbon dioxide in the atmosphere is utilised, which contributes to the reduction of carbon dioxide. Although vegetation has no magic solutions of heavily polluted areas, it improves air quality and reduces CO_2. Indirect benefits urban greenery has on air quality are as follows:

- Reduced air temperature, which prevents inversion phenomena
- Reduced smog
- Contribution to non-scientifically proven positive health effect—by reduction of anions present

The vegetation cools down cities and helps with reduction of number of days with inversion, which have direct impact on public health.

Some types of plants and leaf morphology are more efficient in removing particulate matter than others are. Thus, the selection on the right plants and their incorporation into integrated MUS solution for optimisation of benefits is part of the innovative planning and quantification process.

4.6 Droughts

4.6.1 Why Is Drought a Problem?

The implications of drought on a global scale are enormous. Not only is food production endangered, but also fresh water resources are pressured. The environment is changing. Risk of forest fires is growing and hydropower plants are forced to reduce production or even temporarily shut down. Mostly droughts are caused by lack of participation during winter. The level of moisture in soil is low and the first warm period in spring will lead to drought. This has severe consequences for agriculture and the environment. Biodiversity is also in danger from negative influence of drought. Some of the flora and fauna species are dependent on specific humid ecosystem. With increase in temperature in urban areas, evaporation increases as well. If not controlled, it will lead to increase in desiccation. Most of the vegetation species in urban areas are not resilient to prolonged periods of drought. Thus, irrigation is needed. This comes at the expense of already stressed fresh water supplies. Long dry and warm periods are often followed with heavy rain showers. Dried soil is not capable of absorbing water so the majority of rainwater run-off overflows the sewers. Drought leads to decrease in groundwater table, which in cities can result in damaging building foundations.

4.6.2 What Can Be Changed?

There are no standard solutions available for this problem. It is important that water shortage problems in summer periods be taken into account while designing new or reconstructing existing urban areas. Reducing the amount of impervious surface and/or use of pervious materials that ensure water can infiltrate the soil will help. Generally, there is enough water; it only needs to be managed better. According to MUS solutions, rainwater has to be harvested, stored and used during dry periods together with recycled grey water and eventually black water.

4.7 Urban Agriculture

Urban agriculture is not a new development. After the urban boom following the industrial revolution and during the twentieth century, the lack of food raised was a problem. Today, 15 % of the world's food is produced in cities and it is estimated that it has to be at least doubled within the next 20–30 years. During 2010, a world agriculture report concluded that industrial agriculture is not capable of feeding humanity. One of the biggest problems for agriculture is a lack of space, because of rapid and uncontrolled urbanisation.

Benefits of urban agriculture

- Providing additional and healthier food supply.
- Making material flow more efficient by reducing food mileage.
- Organic waste can be processed (compost) and used directly for food production, thus no transport is needed—leading to energy efficiency.
- Impacts on the thermal absorption capacity of cities and reduction in the urban surface temperature, which has direct influence on heat stress.
- Being part of the city's green lungs and ventilation networks—more green has a positive effect on air quality.
- Increasing pollination and thus biodiversity—particularly without the use of pesticides.
- Having an educational value as well as the important role of social aspects.

Urban agriculture can be practised in various urban spaces (Fig. 18): backyard gardens, front gardens, multifunctional roofs, balconies (vertical gardens), even in cellars (mushrooms). Optimised integration of urban farms into innovative urban planning is art to be mastered.

4.8 Urban Amenity and Blue-Green Corridors Increase Amenity and Urban Health: Job Creation, Reduce Antisocial Behaviour and Crime

During the last decades, more and more data is available on the influence of blue and GI on people's behaviour and health. Most people naturally want to live in greener districts, which will have more functions than conventional GI and will not lose its vitality with first signs of droughts. Different drivers make cities go greener: attract more people, both to work/live or holiday, or increase biodiversity and reduce heat stress and not the least to improve quality of life and appeal of the city. As mentioned before, GI cools down cities, thus providing more pleasant and healthier microclimate. However to be more drought resistant they have to include new (blue) functions.

In addition, the concept of a healthy environment cannot go without GI. It is proven that vegetation helps people to recover faster (Ulrich 1984), thus health costs are significantly lower.

Green areas are places where people can be involved in positive urban regeneration activities (Fig. 19); go for recreation, exercise, to enjoy nature or to find some quiet and peaceful place. Urban vegetation decreases possibility of depression and many other diseases (Maas 2008).

The parks as playgrounds are important for children for their social and communication skills. 'In practice, the lack of sufficient vegetation means, among other

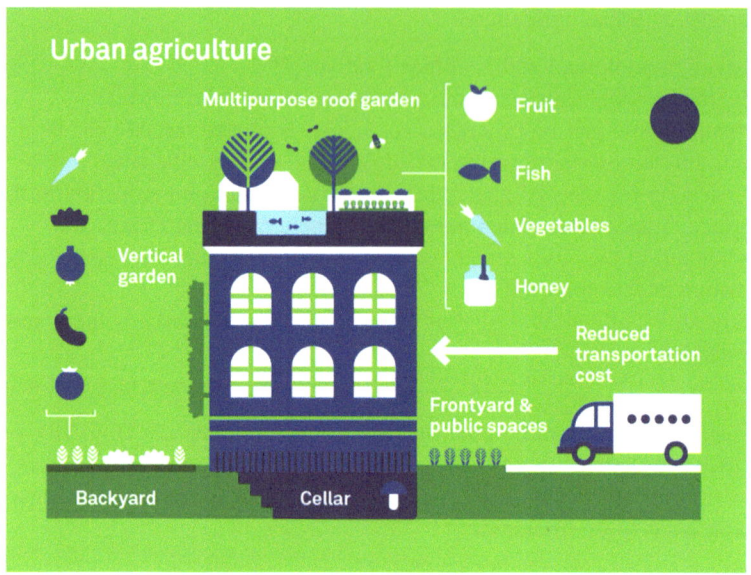

Fig. 19 Spaces for urban farming

things, that there are insufficient possibilities for walking, sports or games and that children are becoming more and more removed from green and nature' (Brosens 2008). Children's physical health will increase—children in green areas are less likely to be overweight (Fig. 20).

Children gain important experiences in nature. Contact with nature helps with self-awareness and increases autonomy. It helps them understand the importance of nature and the environment and the role it plays in human life (Van Den Berg et al.

Fig. 20 Involvement of local stakeholders

2007). Thus, it is important to think about GI as a means to increase human well-being and health. There is research on the influence of small green areas such as neighbourhood vegetation, streets, trees, green roofs, facades or private gardens. Neighbourhood vegetation and parks in cities are seen as valuable. Distances up to 500 m (5-minute walk) can be easily covered on a daily basis, and if that distance is vegetated and pleasant to walk, more people will chose to walk or cycle instead of using a car or public transport. It will allow people to meet, socialise, spend time together without obligations or the need to spend money. Those types of vegetation should be designed to support different age groups.

4.9 How MUS Aligns with and Supports the Concept of Sustainability

The land covered by urban areas is expanding faster than the urban population it supports. From 2000 to 2030, it is expected that the urban population will double (from 2.84 to 4.9 billion people). In the same period, the total land area covered is predicted to triple. By 2050, 70 % of people will live in urban areas.

In the next 40 years, urbanisation will take place in areas with the richest biodiversity. Hence, cities have to become hubs of change. Simon Christmas (Department for Environment, Food & Rural Affairs 2013) explains how we are depleting the environment and placing heavy pressure on natural resources. By 2030, 47 % of the world's population will live in areas of high water stress (OECD 2008) and prime agricultural land will have been built over, with large increases in waste generation, water and soil pollution. Food production by 2050 will have doubled to meet the growing population's dietary needs; Moreover, by 2050, global energy demand will rise by 80 % and demand for water is expected to increase by an additional 55 %. Conventional engineering solutions only displace the effects on ESS elsewhere, rather than solving the problem.

The MUS paradigm offers a new approach for efficient planning and management of the urban environment: one that maximises ecosystem services, minimises negative environmental impacts and increases cities' resilience. It combines natural and technology-based solutions to avoid fragmentation of ecosystems and hence, increase ecological functionality as a whole. The sustainability of MUS solutions can be quantified through indicators of energy efficiency (reduction in demand), water supply (reduce, reuse, recycle), urban health, food security, biodiversity, etc.

It should be noted that MUS is an innovative concept, so there are to date no examples in which the complete MUS methodology is implemented in large scales. However, the examples included in this course, demonstrate how adopting small-scale MUS practises/measures in urban areas can yield major environmental, economic and social benefits.

Attaching a monetary value to ecosystem services can be enormously useful to making a strong case for adopting MUS measures/solutions (see also discussion in Sect. 2). For example, in Cape Town, South Africa, it was calculated that for every unit of currency (ZAR) that municipality spends on the environment, at least 8.3 ZAR of ecosystem goods and services is generated. A similar study on street and park streets was conducted in five US cities. The 2011 assessment of Mayesbrook Park rehabilitation project, East London, showed how MUS solutions could help urban areas to cope with the elevated risk of damage due to climate change, for example increased flooding and higher summer temperatures, while also providing socio-economic gains. The investment of £3.81 million into GI will yield a lifetime benefit-to-cost ratio of approximately 7:1. A study in Toronto showed that by implementing green roofs, urban temperature could be reduced up to 2 °C, resulting in energy savings of $12 million.

5 Conclusions

This volume introduces the methods by which mutual interactions of urban water infrastructure (blue assets) and urban vegetated areas (green assets) are taken into account in the synergy of spatial planning and optimised modelling of ecosystems' performance indicators. This method of planning should make future developments cheaper to build, their users will pay lower utility bills (for water, energy, heating), such developments will be more pleasant to live in and property value would likely be higher. This volume presents the basics of a vision and an understanding of available measures and options for planning of new and retrofitting / redesign of the existing cities based on the MUS concept. No society is going to be rich enough to miss the opportunity to provide integrated multifunctional 'wise men's MUS solutions' presented in this section. To arrive at this vision, we need to keep all stakeholders involved and engaged, since the ambitions and plans are all too often removed from the people living in the area affected. Although the benefits from green and blue infrastructure are fairly well understood, the synergies from their integration are less understood and (much) less applied. Now is a good time to rethink all those benefits and incorporate them into consistent innovative urban planning.

The MUS paradigm has, in this section, been demonstrated to be a highly effective means of mitigating the numerous problems that can arise in towns and cities through adoption of the existing urban planning paradigm. The essential concept is the integration of vegetation into the urban framework to provide multiple benefits without compromising quality of life or the protection of the natural environment. However, to implement MUS solutions successfully, a clear vision and well-defined measures for the desired outcomes are necessary. It is essential, when devising plans, to take into account the needs of the people who are, or will be, resident in the area. The overall objective/goal should be to provide a better place for living. Urban planners are well aware of the benefits that dwellers in urban areas gain from vegetation. For vegetation to continue providing those benefits, it has to have a reliable water supply.

To conclude, it is time to stop seeing the benefits brought by vegetation as removed/separate from conventional urban settlements and to rethink the entire urban planning paradigm with these benefits embedded at their core.

Keywords and Definitions

Multiple-use water services	The project under the European Institute of Innovation and Technology's Climate Knowledge and Innovation Community to 'enhance the synergy of urban blue (water) and green (vegetated) systems and provide effective, multifunctional Blue Green solutions to support urban adaptation to future climatic changes' (http://bgd.org.uk/)
Blue services	Flood protection, water supply (for irrigation, drinking water, land subsidence control), recreational water, and thermal energy collection, transport and storage, space for living and working on or above water, landscaping, habitat for aquatic and terrestrial species and cultural services (physical health, aesthetics, spiritual), climate regulation (equitable climate), detoxification and purification of water (pollution control) and hazard regulation
Climate change	Change in global or regional climate patterns, attributed largely to the increased levels of atmospheric carbon dioxide produced by fossil fuel use
Green services	Parks and recreation grounds, brown-field remediation sites, woodlands, gardens, churchyards and green corridors, trees and standing vegetation (food and timber), wild species diversity, regulating (detoxification) and cultural services (physical health, aesthetics, spiritual) in addition to their natural ability to improve the delivery of climate-related services

Millennium Development Goals	Eight international development goals established following the Millennium Summit of the United Nations, following adoption of the United Nations Millennium Declaration
Pluvial flooding	Flooding that result from rainfall overflow before run-off enters any watercourse or sewer
SUDS	Sustainable Urban Drainage System
UHI	Urban Heat Island
WSUD	Water-Sensitive Urban Design

References

Bahri A (2012) Towards diversification and sustainability. Integrated Urban Water Management (IUWM)

Brosens M, Woestenburg M, De waarde van groen (2008) Ministerie van LNV

Brown R, Keath N, Wong T (2008) Transitioning to water sensitive cities: historical, current and future transition states. In: 11th international conference on urban drainage, Edinburgh, Scotland

CIRIA (1998) CDM regulations: practical guidance for planning supervisors. CIRIA report 173. CIRIA, London (Connection with CIRIA research project 555)

Fratini CF, Geldof GD, Kluck J, Mikkelsen PS (2012) Three points approach (3PA) for urban flood risk management: a tool to support climate change adaptation through trans-disciplinarily and multi-functionality. Urban Water J 9(5):317–331

Luan H (2013) Mitigation of urban heat island with vegetation. MSc dissertation, Imperial College, London

Maas J (2008) Vitamin G: green environments—healthy environments. Ministerie van LNV, Utrecht

Maksimović Č, Stanković S, Liu X, Lalić M (2013) Blue green dream project's solutions for urban areas in the future. In: International science conference reporting for sustainability, Bečići, Montenegro, May 7–10, 2013

Meerow A, Black R (2005) Enviroscaping to conserve energy: a guide to microclimate modification. Department of Environmental Horticulture, Institute of Food and Agricultural Sciences, University of Florida, Gainesville

Novotny V, Elmer V, Furumai H, Kenway S, Philis O (2010) Water and energy framework and footprints for sustainable communities—white paper for the IWA World Water Congress Montreal. International Water Association

Organisation for Economic Co-operation and Development (OECD) (2008) Annual report. OECD, Paris

Rosenthal JK, Crauderueff R, Carte M (2008) Urban heat island mitigation can improve New York City's environment: research on the impacts of mitigation strategies on the urban environment. Sustainable South Bronx working paper. http://www.ssbx.org/index.php?link=37. Accessed 13 April 2014

Ulrich RS (1984) View through a window may influence recovery from surgery. Science 224:420–421

United Nations Environmental Program (UNEP) (2009) 3rd UN World Water Assessment Programme report: water in a changing world

US Environmental Protection Agency (US-EPA) (2012) National water program 2012 strategy: response to climate change

Van Den Berg AE, Haritg T, Staats H et al (2007) Preference for nature in urbanized societies: stress, restoration, and the pursuit of sustainability. J Soc Iss 63(1):79–96

Web Reference

Department for Environment, Food & Rural Affairs (2014) http://www.defra.gov.uk. Accessed 13 April 2014

Chapter 2
What Are the Main Options for Applying the Multiple-Use Water Services Paradigm?

Abstract This brief asks the question, what are the main options to apply the MUS paradigm in urban environments? It breaks down the various components and provides cost–benefit analyses for the various components along with challenges and considerations for both the short and long terms. The brief includes a section on the MUS approach and a means to calculate the value of MUS systems, as well as provides tools and resources to support urban blue-green design. Comprised of actual and potential options for decision makers and policy makers to integrate blue and green measures that target the optimal synergies between interventions and techniques with the purpose of delivering multiple benefits, reproducing the natural pre-development process to the best possible degree and boosting ecosystem services.

Keywords ESS · Green roof · Green walls · Infiltration trench · MUS · MUSIC · Permeable pavement · Rain gardens · Retention ponds · Swales · SWMM · Urban agriculture · Urban water management · UWOT · WASP

1 Introduction

An overview of main concepts and techniques for achieving the envisaged MUS synergy in urban design and management is presented in Fig. 1 and analysed in detail in the following sections.

2 Wastewater Reuse and Recycling

Wastewater reuse—recycling can be generally defined as the use of treated water for several purposes such as toilet flushing, landscape irrigation, groundwater recharge, etc. Wastewater is a general term for used water that its quality has been affected from human activities, residential, commercial or industrial. Wastewater is usually discharged through centralised sewer systems to Wastewater Treatment Plants for removal or reduction of hazardous substances. Alternatively, wastewater

© The Author(s) 2015   27
Č. Maksimović et al., *Rethinking Infrastructure Design for Multi-Use
Water Services*, SpringerBriefs in Environmental Science,
DOI 10.1007/978-3-319-06275-4_2

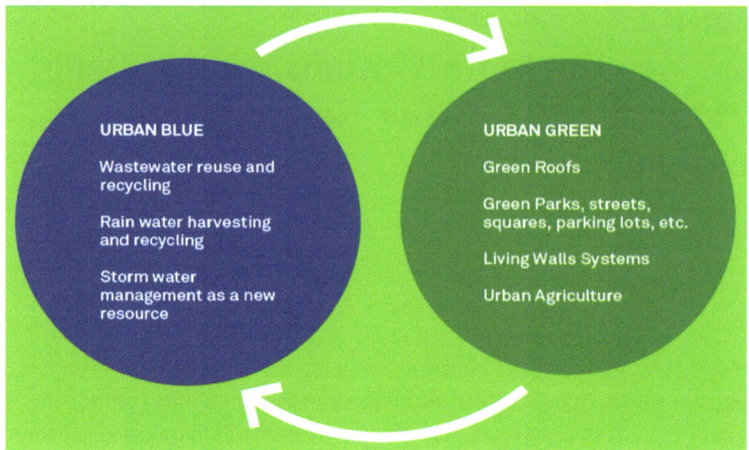

Fig. 1 The main options for multiple-use water services paradigm

can be treated on-site. The term 'reused' water is often used similarly with 'recycling' water or 'reclaimed' water.

Residential wastewater, based on the source and level of contamination can be further classified as black (highly polluted) and grey. Black water, usually referred to as wastewater, is highly contaminated water coming from toilets and urinals. Grey water, on the other hand, is less polluted water discharged from sinks, showers and bathtubs, washing machines and drinking fountains. When reusing grey water, water discharged from kitchen sinks and dishwashers is generally excluded due to higher levels of contamination coming from food residues and animal products. MUS mainly focuses on the on-site use of reclaimed grey water used for multiple non-potable purposes such as indoor and outdoor plant irrigation, including green roofs and walls, and toilet flushing.

2.1 What Are the Main Components and Costs Related to Grey Water Reuse?

Grey water reuse systems can vary significantly from simple, low-cost appliances that harvest grey water and convey it for direct use, e.g. in toilets and gardens, to composite systems integrating specialised treatment processes.

Cost and energy required can also vary, mainly increasing as more and better treatment is involved. Grey water reuse systems are more suitable for new-built developments, as retrofitting existing systems can be more expensive, but they can be incorporated while renovation and plumbing replacement activities occur (CGBC 2011).

Lookout Grey water policies and proper regulation are of key importance to reinforce public acceptance and awareness, economic viability and implementation of grey water reuse practices, aiming at reducing water demand and improving water sustainability (Yu et al. 2013). Economic incentives, such as subsidies, provided by water utilities can be valuable tools to promote grey water reuse technologies. Tucson Water, in the city of Tucson, Arizona, through the Gray Water Rebate Program offers rebates up to $1,000 per household for the installation of grey water irrigation systems for both retrofits and new buildings (www.tucsonaz.gov/water/rebate).

2.2 Why Grey Water Recycling?

- Reduction of water demand from public water supply for non-potable uses, leading both to lower household water bills and wider community benefits.
- Reduction of effluent discharge and thus energy reduction for wastewater treatment.
- Household water savings can reach a level of 50 % through grey water reuse for toilet flushing and garden irrigation (Maimon et al. 2010).

3 Urban Green Spaces

Green spaces in the cities include private gardens, parks, green parking lots, squares and streets, community forests, etc.

3.1 What Is the Cost-Benefit?

Planting and maintaining trees and vegetation can be costly. The main costs associated include initial planting and ongoing maintenance, such as for irrigation, pruning and pest control, administration, etc. Nevertheless, the benefits derived (direct and indirect) can exceed the overall cost.

Facts A study on the functioning and value of street and park trees in five US cities from different States (McPherson et al. 2005) showed that for every dollar invested in tree planting and ongoing maintenance, benefits returned annually ranged from $1.37 to $3.09. More specifically annual costs ranged from around $15–$65 per tree, while total revenues including energy savings, atmospheric CO_2 and storm water run-off reduction, air quality, aesthetics and other benefits were about $31–$89 per tree. Regarding costs, pruning was

Table 1 Benefits of rainwater harvesting

Short-term benefits	Long-term benefits
Meet water demand when no other water sources are available	Reduced storm water run-off leading to lower energy consumption for storm water treatment
Reduction of water demand	Use of harvested rainwater for aquifer recharge and increase of depleted groundwater table
High collection and distribution efficiency	Reduction of diffuse pollution resulting in improvement of aquatic ecosystems
Self-sufficiency (less dependency on distant watercourses)	Potential for lower consumer water bills
Reduction of flood risk (reduction of economic losses)	Greater flexibility of a decentralised system consisting of numerous water resource points in case of a natural disaster rather than a centralised water supply system that may collapse or go out of order
Enhance rational utilisation of water through decentralised systems	

found to be the most expensive practice, with related costs to be around 25–40 % of the total annual cost, followed by administration and inspection costs (8–35 % of total annual costs), while tree planting cost was estimated to be 2–15 % of total annual urban forestry expenditures.

4 Rainwater Harvesting

Rainwater harvesting (RWH) is a decentralised technique of collecting and storing rainwater for later use at or near the point where water is needed or used providing multiple benefits (Table 1). Depending on scale, requirements and purpose, RWH systems can range from low storage capacity (50–100 gallon) systems (e.g. rain barrels) to larger systems (1,000–100,000 gallons) (US EPA 2013). Rain barrels can be easily placed outside buildings, with no connections to internal or external plumbing, where rooftop run-off from downspouts is captured for later use mainly for outdoor purposes, such as car washing and irrigation. Higher volume systems (e.g. cisterns) collect storm water from roofs and other surfaces (e.g. parking lots, terraces), and after quality treatment provide water to a distribution system. Harvested water can be used outdoors (e.g. landscape irrigation, fountains) or indoors (e.g. toilet flushing, clothes washing).

A typical RWH system is mainly comprised of the catchment area upon which the rain falls; storage tanks and cisterns; gutters and downspouts to transfer rainwater from the catchment area to the storage system; a filtering system to remove debris, solids and other materials; a monitoring system (e.g. for monitoring the water level inside the tank) and a system to convey water for further use (e.g. gravity system or pumps). The main issues emerging when constructing an RWH

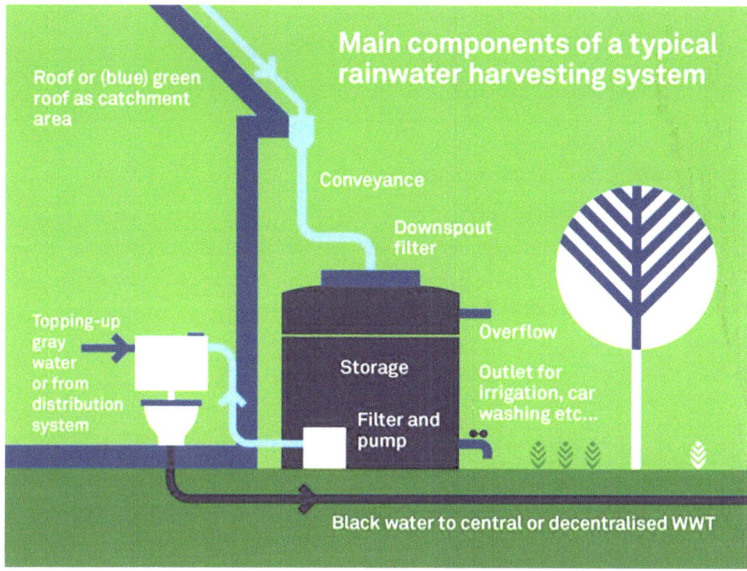

Fig. 2 Main components of a typical rainwater harvesting system

system are the availability and cost of materials; labour cost; space availability; local expertise for the construction of the system; consideration of local traditions on water storage; climate conditions and catchment characteristics (Fig. 2).

4.1 New Build Versus Retrofit

Rainwater harvesting can be applied in most buildings but it is more suitable for new constructions due to the fact that the installation of an underground tank could be very expensive and may involve re-routing of some services as well as fitting a tank and filter in an existing drainage scheme will involve changes in the pipe system (CGBC 2011).

4.2 What Is the Cost?

The cost of an RWH system is site-specific and varies significantly depending on size, type and complexity of the system. The cost of a rain barrel can differ based on material and size, with a typical 50-gallon, plastic rain barrel to cost around $70 (US EPA 2013). For larger systems that do not have significant filtration or

Table 2 Maintenance activities and costs associated with cisterns

	Months between events	Cost per event	Total cost per year
Routine maintenance activities			
Inspection, reporting and information management	6	$130	$260
Roof washing, cleaning inflow filters	6	$240	$480
Corrective and infrequent maintenance activities (*unplanned and/or >3 years between events*)			
Intermittent system maintenance (system flush, debris/sediment removal from tank)	3	$390	$130
Pump replacement	5	$989	$198

Source WERF (2009)

distribution requirements, storage is usually the most expensive element. A cistern can cost around $1.50–$3.00 per gallon of storage. The more complex the system, the higher the capital cost. Filtering, pumping, distribution and treatment systems, plumbing and drainage connections, excavations, installations and other elements can have an additional cost of around $2–$5 per gallon (Table 2).

5 Green Roofs and Green Walls

5.1 What Are Green Roofs?

A green roof, also known as eco-roof, living or vegetated roof, is a roof of a building that is entirely or to an extent covered with vegetation planted over a waterproofing membrane. Green roofs can be categorised, depending on the depth of planting medium and level of maintenance they need, as extensive, semi-intensive and intensive (Fig. 3).

A typical green roof is a complex system of several layers of materials to attain waterproofing and to remove water from the roof deck (Tolderlund 2010; Fig. 4).

5.2 What Are the Risks and Costs Associated with Green Roofs?

In case of bad construction or inefficient maintenance, a green roof may face the risk of leakage and damage or even collapse, or may fail to deliver the desired energy efficiency levels. The main factors affecting the cost of green roofs include the type of structure (extensive or intensive), types of vegetation, irrigation systems, accessibility, retrofit or new development. Generally, an intensive green roof has a higher capital and maintenance cost than an extensive green roof, as well as

	Extensive	Semi-Intensive	Intensive
	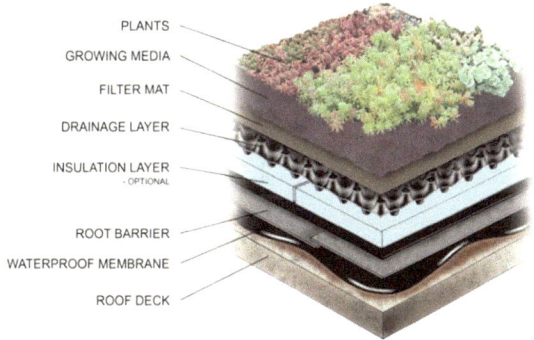		
Depth of growing medium	3 - 5 inches	5 - 7 inches	7 - 24+ inches
Weight max.	15- 25 lbs/ft²	25- 40 lbs/ft²	35 - 80+ lbs/ft²
Plants	sedums, small grasses, herbs and flowering herbaceous plants	selected perennials, sedums, ornamental grasses, herbs and small shrubs	perennials, lawn, shrubs and small trees, rooftop farming
Irrigation	Not recommended	Occasional irrigation	Advanced irrigation systems
Maintenance	low	medium	high
Use	Living machine	Diversity, habitat	Garden, park
Costs	low	medium	high

Fig. 3 Types of green roofs (*Source* Adapted from http://www.greenrooftechnology.com)

Fig. 4 Typical green roof
layers (Photo courtesy
American Hydrotech, Inc.
Source Tolderlund (2010))

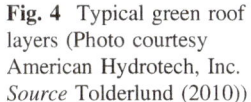

PLANTS

GROWING MEDIA

FILTER MAT

DRAINAGE LAYER

INSULATION LAYER
- OPTIONAL

ROOT BARRIER

WATERPROOF MEMBRANE

ROOF DECK

installing a green roof in a new construction costs less than retrofitting (Peck and Kuhn 2003). The installation of an extensive green roof with root repellent/ waterproof membranes may cost around $10–$24 US/sq.ft. (GRHC 2014). Green roofs are more expensive to construct than conventional roofs but they can be more cost-effective in the end due to extending the life span of the roof membrane and leading to significant energy savings from heating and cooling.

Fig. 5 A green wall (*left*) and a green façade (*right*) in Melbourne (*Source* DEPI (2014))

Facts A 2006 study conducted by the University of Michigan compared the costs and benefits, including storm water management and public health improvement, of a 21,000 sq.ft. green roof over a conventional roof. The results showed that the installation of the green roof would cost $464,000, while the conventional one $335,000 (in 2006 values). However, it was estimated that the green roof would save more than $200,000 during its lifetime, with two-thirds of that savings resulting from reduction in energy needs (Clark et al. 2008a, b).

5.3 What Are Green Walls?

Green walls, also known as living walls, bio- or eco-walls are vertical plants either grown on freestanding structures or attached to interior or exterior walls (DEPI 2014). They are composite systems incorporating plants, growing medium, drainage, irrigation and often fertilisation. They can retain a great variety of vegetation depending on local climatic conditions. In green walls, the whole structure is plated compared to **green façades** where climbing plants are used, which are either rooted in the ground at the bottom of the structure or are planted in boxes at different levels and cover a part or the entire surface of a building. However, green wall is a general term used for interior or exterior vertical vegetated surfaces (GRHC 2014). They are usually designed for aesthetic purposes but they can provide additional benefits such as enhanced interior and exterior air quality, thermal insulation and higher property values (Fig. 5).

5.4 Green Walls Used for Wastewater Treatment

Water recycling systems can be linked to green walls, pumping the captured grey water (or even collecting storm water), which then passes through gravel filters and marine plants for treatment. The effluent is then directed to a storage tank for domestic use or to public water treatment plants (GRHC 2014).

Showcase: EDITT Tower in Singapore

Photorealistic image of the EDITT Tower in Singapore, designed with a grey water filtration system that includes water purification, water and wastewater recycling. Only 45 % of building water demand is supplied through the public mains.

Source TR Hamzah & Yeang Sdn Bhd (http://www.trhamzahyeang.com/project/skyscrapers/edit-tower01.html)

5.5 What to Consider Before Applying Green Walls and Façades?

While selecting a green wall system one must consider installation and maintenance cost and structure requirements, climate conditions, lighting, types of plants and quality, functionality and lifespan (DEPI 2014). A successful construction must serve its design purpose, require low maintenance, effectively support the selected vegetation and have a long lifespan.

5.6 Why Green Roof and Green Wall Systems?

Mitigation of Urban Heat Island Effect—Reduced air temperature through shading, evaporation and light absorption provided by plants.

Enhanced air quality by capturing airborne pollution, harmful gases and volatile organic compounds and providing thermal insulation inside buildings resulting in reduced energy demand for heating and thus less CO_2 released into the air.

Increased biodiversity in an urban environment—Green roofs and walls can sustain a range of vegetation and serve as habitat and nesting place for different bird species.

Local job creation in the fields of design, manufacturing, installation and maintenance.

Enhanced aesthetics, amenities and recreational green spaces (e.g. community gardens, playgrounds in green roofs) and *increased property values of buildings.*

Storm water retention and water filtration through green roofs—Green roofs can return 50 % of annual precipitation back to the atmosphere through retention and evapotranspiration (Berghage et al. 2009). In addition to reducing the volume of storm water run-off, a green roof can successfully delay the time to peak, leading to less stress on sewer systems at peak flow periods.

Thermal insulation and energy savings—In summer, an extensive green roof can reduce daily energy demand for air conditioning during summer by 75 % (Liu and Baskaran 2003).

Noise reduction—Vegetated vertical and horizontal surfaces can block high-frequency sounds and when combined with a substrate or growing medium can block low-frequency sounds. Extensive and intensive green roofs can reduce sounds from outside the building by 40 and by 46–50 decibels, respectively (Peck et al. 1999).

Fire Retardation—Green roofs are found to have better fire resistance values compared to conventional roofs (Köhler 2004).

Urban agriculture—With specific design green roofs and walls are suitable for growing fruits, vegetables and herbs.

Extended roof life—Green roof systems provide protection to roofing membranes from the effects of UV light, mechanical damage, high thermal temperature fluctuations therefore leading to a longer lifespan.

Reduction of electromagnetic radiation—Green roofs can reduce electromagnetic radiation penetration by 99.4 % (Herman 2003).

6 Urban Agriculture

Urban agriculture is generally the practice of cultivating crops for food in cities. Growing fruits, vegetables and herbs in cities can be combined with other green infrastructure (green roofs and walls) and decentralised water management techniques (RWH) (Fig. 6).

Fig. 6 Blue green urban agriculture

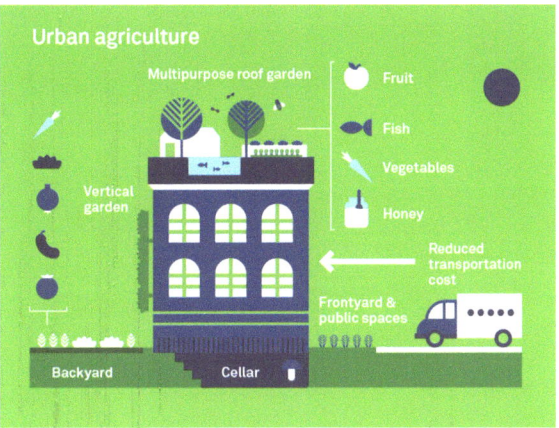

6.1 What Are the Challenges Regarding Water Conservation?

Cultivating crops for food often requires significant amount of water for irrigation, which could be an 'extra burden' for water demand in urban areas. Moreover, municipal water supplies are usually more expensive and energy consuming than agricultural water supplies, as municipal water must go through particular treatment to meet drinking water standards (Nolasco 2011). Thus, it is of key importance that water-use efficiency and conservation practices are incorporated into urban agriculture. Such techniques could include rainwater harvesting, grey water harvesting, physical water retention methods, drip irrigation, etc.

Showcase Zuidpark, The largest rooftop garden in Europe

In 2012, the conventional rooftop of Zuidpark, a former administrative building in an industrial zone of Amsterdam, was redeveloped to host a 3,000 m² garden, so far the largest in Europe, where organic vegetables, herbs and flowers are cultivated (IGRA 2012). The roof garden substrate was specifically developed to be used for urban agriculture. Unlike common practices of urban agriculture, agricultural products are not for sale but are served to the building's common restaurant. The rooftop farm is a place where people meet, rest, have lunch and can even take gardening lessons. Most of all, Zuidpark's farm roof represents a great example of creative thinking.

Source Helga Fassbinde (http://www.biotope-city.net)

7 Decentralised Systems to Manage and Reuse Storm Water Run-off On-site

7.1 What is Decentralised Storm Water Management?

Storm water management involves practices that improve the quality and reduce the quantity of storm water run-off, in order to prevent or mitigate flooding, waterways contamination and negative-related consequences, such as infrastructure damage, bank alterations and erosion, habitat destruction and quality degradation of streams, rivers and coastal waters.

Traditionally, storm water run-off from streets, roofs, pavements and other impervious surfaces in urban areas is collected through pipes and sewer systems and conveyed quickly offsite, where it is either discarded straight away into a water receiver, or it is first treated by a Wastewater Treatment Plant. Also known as a 'drained city' approach. Currently, urban water management puts emphasis on the decentralised storm water and rainwater management such as Low Impact Development (LID) for USA, Decentralized Urban Design (DUD) for Germany, Water-Sensitive Urban Design (WSUD) of Australia, Sound Water Cycle on National Planning (SWCNP) for Japan and Smart Watergy City (SWC) for South Korea, and have similar concepts with on-site rainwater and storm water management and source control (Table 3).

The MUS alternative, offers a distributed approach involving the implementation of different decentralised on-site storm water management techniques, trying to mimic the natural drainage process. These decentralised systems select the attenuation (temporary storing and release at a later stage), infiltration to the ground, or conveyance (slow transport) of the urban run-off. In addition, they aim at filtering out pollutants and allowing sediment settlement. They integrate the normative

Table 3 Decentralised storm and rainwater management approaches among countries

Classification	Concept	Characteristics
South Korea	Smart Watergy City, U-Eco City (SWC)	Water management based on ubiquitous Construct of ecosystem using green energy and technology Watergy: water, energy and ecology
USA	Low impact development (LID)	Management of pollution sources and rainwater management based on green land Best management practices (BMPs) Water quality capture volume (WQCV) Green infrastructure (GI) Smart water grid
Australia	Water-sensitive urban design (WSUD)	Rainwater management adaptable to climate change Management and using of storm water run-off
Germany	Decentralised urban design (DUD)	Decentralised rainwater management by arcology Management and using of storm water run-off
Japan	Sound water cycle on national planning (SWCNP)	Sound water cycle by rainwater management Reduction of storm water run-off Detention and infiltration in watershed

Source UN ESCAP (2012)

values of environmental protection and restoration, water supply reliability, flood control, public health, amenity and leisure, energy consumption reduction, climate change adaptation and economic viability.

7.2 What Are the Typical Schemes and Techniques?

The decentralised storm water management schemes use a combination of processes, mechanism and components to deliver their expected benefits. The processes involved in these schemes can be broadly classified as source control, swales and conveyance channels, filtration, infiltration, retention and detention, wetlands, inlets/outlets and control structures.[1] The MUS concept incorporates these distributed storm water management techniques with a shift to the green infrastructure approach (US EPA 2008), to be implemented in different scales: site-specific, neighbourhood and regional. Wide-scale design and implementation of combined Green Storm water Infrastructure tools such as rain gardens, infiltration systems, constructed wetlands, vegetated swales, etc. can provide numerous benefits and support a sustainable Blue and Green urban environment (Table 4).

---

[1] For more information, visit http://www.susdrain.org.

Table 4 Decentralised storm water management techniques

Categories	Techniques/measures
Source control	Green roofs, rainwater harvesting, permeable paving and other permeable surfaces
Swales and conveyance channels	Swales, channels and rills
Filtration	Filter strips, filter trenches, bioretention areas
Infiltration	Soakaways, infiltration trenches, infiltration basins and rain gardens
Retention and detention	Detention basins and retention ponds
Wetlands	Wetlands

Source www.susdrain.org

What Are Rain Gardens?

Rain gardens are shallow-planted depressions designed to receive rainwater from hard surfaces such as roofs, paved areas or roads. The excess run-off infiltrates into the soil, reducing peak flow on site and recharges groundwater. The soil layers underneath also assist in the removal of pollution, such as nitrogen, phosphorus and fertilisers, which are washed off from hard surfaces. The plants in the rain garden help to further filter out pollution. Rain gardens can be applied at a variety of scales and are self-sufficient compared to regular gardens as they use storm water directly, thus resulting in reduced domestic water use for gardening.

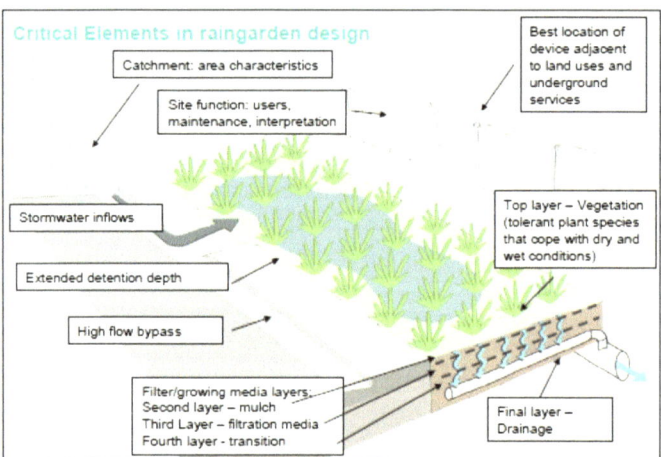

Raingarden design principles. *Source* Clear Water (2012a); http://www.clearwater.asn.au/

Opportunities for Retrofitting: Rain gardens can be constructed at a low cost in new or existing sites. They could be easily retrofitted to existing domestic houses, commercial and industrial buildings with downpipes connected to subsurface water drains.

What Is Permeable Paving?

Permeable pavement is a method of paving that allows water to infiltrate into the ground as it falls rather than running off into piped storm water drainage system. Porous pavements are mostly suitable to be implemented on light traffic loads such as streets with low traffic volumes, parking lots, private driveways, pedestrian paths or footpaths, public squares, etc. They are most effective when used in conjunction with other measures such as vegetable swales, cisterns, etc.

Source Water Sensitive Urban Design in Sydney http://www.wsud.org/

Common Types:

- Porous asphalt is the same as regular asphalt except it is manufactured with the fine material omitted, leaving voids that allows water to infiltrate.
- Concrete, ceramic or plastic pavers—Designed to leave gaps between allowing run-off penetration.
- Grid systems or open cell pavers are made from plastic or concrete grid filled with soil or aggregate so that water can percolate through

Source University of Maryland Extension (2011)

Design Characteristics: There are several options for the design and construction of systems. After infiltrating through the pavement surface, the water can be stored in a sink tank or in plastic cellular systems.

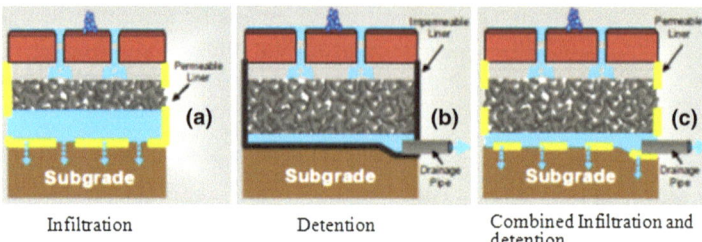

| Infiltration | Detention | Combined Infiltration and detention |

The associated costs for impervious pavements are influenced by the type of material, the preparation of the site, the installation and the maintenance of the system. Installation and maintenance is likely to be more expensive than the construction of conventional impervious surfaces. Typical construction costs vary from $5—10/sq.ft. (Clark et al. 2008a, b).

What Are Swales?

Swales are shallow, broad and vegetated channels filled with porous filter media to provide on-site treatment of storm water run-off. Storm water is directed and collected into the shallow depressed area and slowly filters through the vegetated soil media where pollution is removed through physical and biological processes. Water then passes through a transition layer and finally drains into a drainage layer. Depending on the design, treated storm water is usually collected through a piping system inside the draining layer and led downstream to waterways or storage systems, or can infiltrate through the underlying soils. Swales should also contain an overflow or inlet for flood events. Swales need regular maintenance.

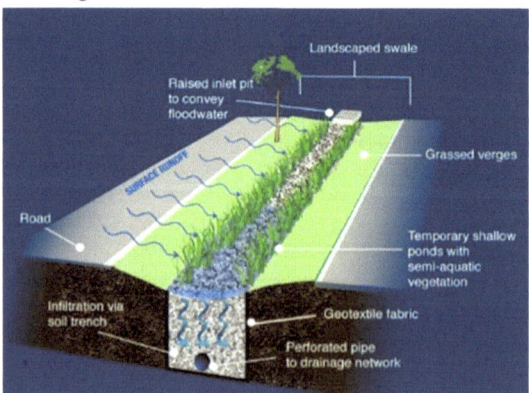

Schematic of a typical bioretention swale (*Source* FAWB, Facility for Advancing Water Biofiltration (2008))

What Are Infiltration Trenches?

An infiltration trench is an excavation filled with permeable material, such as rock and gravel, which is used to capture, treat, store and infiltrate storm water, enhancing the natural capacity of the ground to store and drain. Infiltration trenches allow water to infiltrate into the soil from the bottom and sides of the trench. The treatment procedure involves retention of sediments, nutrients, dissolved heavy metals and other toxic substances. They can be constructed at open spaces, such as parking lots and streets, as a simple trench system or combined with other filtering systems, such as grassed swales and vegetated filter strips to increase pollution removal.

Main limitations regarding this technique concern the high clogging potential, the regular maintenance needed to remove retained pollutants and preserve efficiency and the risk of groundwater contamination if soils are coarse.

Source Melbourne water

http://www.melbournewater.com.au/Planning-and-building/Stormwater-management/

What Are Retention Ponds? Retention ponds can provide both storm water attenuation and treatment, while supporting emergent and submerged aquatic vegetation along their shoreline. Run-off is detained in the pool, while the retention time promotes pollutant removal through sedimentation and biological uptake mechanisms. Maintenance requires removal of debris and litter, cleaning of the inlet, sediment removal and vegetation management.

The need for adequate surface may constrain the construction of retention ponds in highly dense urban areas, while if the inflow is limited (due to the small number of storm events) and combined with poor maintenance anaerobic conditions may occur and consequent health risk.

Source www.susdrain.org

What Is a Constructed Wetland? Constructed wetlands are artificial treatment systems that mimic the physical, chemical and biological functioning of natural wetlands in order to remove pollutants from storm water. The main processes involved are physical detention and filtration of suspended solids and dissolved pollutants as well as biological and chemical uptake by the wetland vegetation. According to US EPA, constructed wetlands are among the most effective measures to remove contaminants from storm water and present high range of applicability, excluding highly urbanised areas and arid climates.

Moreover, they can increase aesthetics and provide habitat to several ecosystems. Careful consideration must be integrated into the design before constructing a wetland to manage significant issues such as the necessity of a large open space, the undesired presence of mosquitos and possible disturbance of the natural environment.

Storm water Wetland in Philadelphia (*Source* http://www.phillywatersheds. org)

8 Integrating Multiple-Use Perspectives

The integration of multiple-use perspective targets the optimal synergies between the aforementioned interventions and techniques with the purpose of delivering multiple benefits, reproducing the natural pre-development process to the best possible degree and boosting the ecosystem services. For example, combining green roofs with urban agriculture and rainwater harvesting can provide storage and peak flow reduction while increasing food provision, aesthetics and leisure, reducing energy consumption and mitigating the heat island effect. Key to a concrete integration is planning and governance. To achieve the desired results urban planning must endorse the blue-green thinking paradigm, while an appropriate institutional setting must be in place to act as an enabling support environment.

Securing financial resources and providing incentives (including subsidies) are also important factors for the uptake and expansion of these techniques. Design must consider the vertical integration of measures and their horizontal application in all possible scales: single dwellings, residential multi-units, public and commercial buildings, streetscapes, blocks, etc.

Some example applications of combined blue-green interventions are provided below, followed by a summary overview of the benefits (Table 5) and indicative costs (Table 6) of the blue-green techniques described before.

Example 1

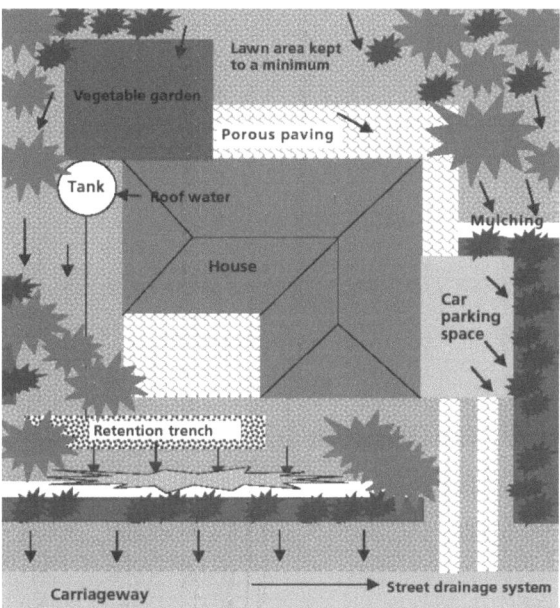

Possible overall strategy for a typical suburban home. A rainwater tank supplies rainwater for toilet flushing, washing machine, and for outdoor use whilst water efficient fittings reduce main water consumption elsewhere. During prolonged or heavy rain, water overflows from the rainwater tank to a retention trench. Storm water run-off from paths, driveways and lawns is directed to garden areas. Excess run-off from impervious surfaces is directed to the retention trench, or overflows to the street drainage system.

Source Hobart City Council (2006)

Table 5 Overview of the benefits of the various Multiple-Use perspectives and techniques

Blue and green measures	Water				Natural hazards		Environment			
	Water supply/reduction of water demand	Reduction of storm water runoff	Restoration of water cycle/aquifer recharge	Reduction diffuse pollution/water treatment	Flood risk reduction	Disaster management (e.g. firefighting)	Mitigation of urban heat island effect/improvement of microclimate	CO₂ reduction/improved air quality	Reduction of noise	Increased biodiversity
Green roofs	(x)	x	x	x	x		x	x	x	x
Living walls, green facades		x		x			x	x	x	x
Green spaces (gardens, parks, lots, squares, streets, etc.)		x	x	x	x		x	x	x	x
Permeable paving	(x)	x	x	x	x					
Swales/bioswales		x	x	x	x	(x)		x		x
Filter strips/trenches		x	(x)	x	x					x
Rain gardens	x	x	x	x	x		x	x		x
Retention/detention ponds	x	x	x	x	x	x	x	x	x	x
Constructed/artificial wetlands	x	x	x	x	x	x	x	x	x	x

(continued)

Table 5 (continued)

Blue and green measures	Water				Natural hazards		Environment			
	Water supply/ reduction of water demand	Reduction of storm water runoff	Restoration of water cycle/ aquifer recharge	Reduction diffuse pollution/ water treatment	Flood risk reduction	Disaster management (e.g. firefighting)	Mitigation of urban heat island effect/ improvement of microclimate	CO$_2$ reduction/ improved air quality	Reduction of noise	Increased biodiversity
Rainwater harvesting	x	x	x	(x)	x	x				
Wastewater reuse and recycling	x	x	(x)	x	x	x		x		
Urban agriculture	(x)	x	x	x	x		x	x		

Blue and green measures	Socio-economic						Well-being			Education, RTD	
	Energy saving	Lowering of water and/or energy bills	Food supply	Self sufficiency	Local job creation	Poverty alleviation	Amenity and aesthetic improvement	Leisure	Health	Education and awareness	Technology development
Green roofs	x	x	(x)		x		x	x	x	x	x
Living walls, green facades	x	x	(x)		x		x		x	x	
Green spaces (gardens, parks, lots, squares, streets, etc.)			X		x	x	x	x	x	x	
Permeable paving					x		x				x
Swales/bioswales					x		x				
Filter strips/ trenches					x		x				

(continued)

Table 5 (continued)

Blue and green measures	Socio-economic						Well-being			Education, RTD	
	Energy saving	Lowering of water and/or energy bills	Food supply	Self sufficiency	Local job creation	Poverty alleviation	Amenity and aesthetic improvement	Leisure	Health	Education and awareness	Technology development
Rain gardens	x	x	X	x	x		x		x	x	
Retention/detention ponds	x				x		x	x	x	x	
Constructed/artificial wetlands	x				x	x	x	x	x	x	x
Rainwater harvesting	x	x	(x)	x	x	x	x		x	x	x
Wastewater reuse and recycling	x	x	(x)	x	x	(x)				x	x
Urban agriculture	x		X	x	x	x	x		x	x	

Table 6 Indicative costs of multiple-use perspectives and techniques

BG measures	Installation cost	Maintenance cost
Green roofs	10$/sq. ft. (extensive green roof) and 25$/sq. ft. (Intensive green roof) (US EPA 2014a, b)	0.75–1.50/sq. ft. annually (US EPA 2014a, b)
Living walls	The cost of materials of a living wall ranges from $60 to $90 per sq. ft. However, the overall cost (plants, soil irrigation, installation) of construction could reach double this amount. (Continuing Education Centre 2014) *Showcase: Large-scale outdoor green wall hydroponic, 2009 The wall is 206 sq. m. with a total cost of $350,000 (158 $/sq. ft.)* (DEPI 2014)	Pruning and panels adjustment :14.41€ ($19.81)/m^2/year (Perini and Rosasco 2013) Irrigation: 0.96€ ($1.32)/m^2/year Panels replacement (5 %): 6.05€ ($8.32)/m^2/year Plant species replacement (10 %): 2.75 € ($3.78)/m^2/year Pipes replacement (irrigation system): 2.85€ ($3.92)/m^2/year Total Maintenance cost : 27€ ($37.11)/m^2/year (2.5€ ($3.4)/sq. ft./year))
Permeable paving	1. Porous Concrete: $2.00 to $6.50/sq. ft. 2. Porous Asphalt: $0.50 to $1.00/sq. ft. 3. Interlocking Pavers: $5.00 to $10.00/sq. ft. (University of Maryland Extension 2011)	Annual maintenance costs about 1–2 % of the construction cost (Prince George's County, Maryland 2014)
Swales/ bioswales	Swales: 15–20$/m^2 (Fletcher et al. 2003) Swale bioretention systems: $100–120/linear metre including vegetation (for this system the filter zone has a width of 1 m and the swale has a top width of 3–4 m) (Leinster 2004)	$2.50—Grass swale ($/m^2/yr) $9.00 Vegetated swales ($/m^2/yr) (initial) $1.50 Vegetated swales ($/m^2/yr) (after 5 yrs) (Lloyd et al. 2002)
Buffer/filter strips	$10–$15/sq. meter—Sydney Grass buffer strip $20–$50/sq. meter—Native grasses and shrubs (URS 2003)	Typical maintenance costs are about $350/acre/year (US EPA 2014b)
Rain gardens	Cost will vary depending on the garden's size and the types of vegetation used; however, professional installation of a rain garden typically costs $10–$12/sq. ft. (Charles River Watershed Association 2008)	The Typical Annual Maintenance cost is estimated as 5–7 % of the construction cost. Maintenance costs are likely to be higher in the first few years due to the intensive effort needed to establish the system (Environmental Protection Agency, Victoria 2008)
Retention/ detention ponds	Typical construction costs in 2004 dollars range from approximately $25,000 to $50,000 per acre-foot of storage. (Pennsylvania Department of Environmental Protection 2006)	Annual cost of maintenance (especially sediment and vegetation removal) estimated at 3–5 % of construction costs (Pennsylvania Department of Environmental Protection 2006)

(continued)

Table 6 (continued)

BG measures	Installation cost	Maintenance cost
Constructed/ artificial wetlands	Small-scale wetland with an inlet pond, macrophyte zone, bypass weir and channel: $90–$100/m^2. Larger-scale wetland to treat recirculated lake water: $65/m^2 (Leinster 2004)	Wetlands typically cost 2–6 % of the construction cost to maintain each year. Smaller wetlands are cheaper to maintain (Environmental Protection Agency, Victoria 2008)
Rainwater harvesting	The capital cost of an RWH system can range from $1.50 to $3.00 per gallon of storage (for simple systems) to $3.5–$8 per gallon for more sophisticated systems (US EPA 2013)	Total Annual cost of primary routine maintenance and corrective activities associated with cisterns was estimated at around $1000 (WERF 2009; US EPA 2013)
Urban green spaces	A study on the cost-benefit performance of street and park trees in five US cities from different States (McPherson et al. 2005) estimated total annual municipal expenditures for tree planting and ongoing maintenance to range from $15 to $65 per tree (McPherson et al. 2005)	

Example 2

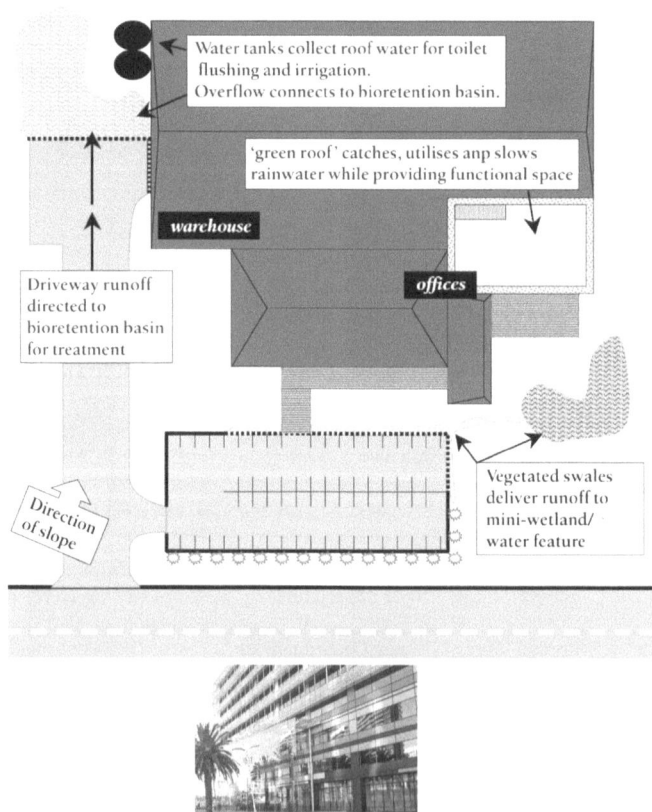

Diagram of a possible industrial site layout incorporating a mini-wetland landscape feature, green roof, vegetated swales delivering car park run-off to the mini-wetland, rainwater harvesting for non-potable uses and a bioretention rain garden to collect driveway run-off and water tank overflow.

Source Hobart City Council (2006)

Example 3

Royal Park Storm water Harvesting Project—City of Melbourne Royal Park, Melbourne, Victoria

The Trin Warren Tam-borne urban wetland (also known as the Royal Park Wetlands) was constructed to treat storm water and run-off from the roads, to provide a habitat for wildlife as well as for supplying treated water to the residents of Melbourne. The wetland was engineered to have two linked ponds into which storm water is diverted to be treated by native Australian plants and other biological processes.

The treated water was then diverted into a storage wetland passing through an ultraviolet disinfection system in order to be utilised for city purpose irrigation. The construction also contains a 6-million-litre groundwater storage facility with two distribution tanks. The wetland has provided home for more than 270 species of birds and the White's Shink lizard.

Source Clear Water (2012b).

9 The MUS Approach in a Rural Context

In many areas around the world where centralised water services are absent, populations living in poverty have limited or no access to clean water for the satisfaction of various needs, ranging from domestic uses, such as drinking, cooking, hygiene and sanitation, to productive activities (e.g. irrigation, livestock production and small-scale enterprises such as brick making and food processing). Existing applications in low-income countries rely on the single-use approach where planning, investment and management of water services targets a single use such as drinking or irrigation with possible negative impacts in human health and sustainability as in the end people use the supplied water for multiple purposes (Renwick et al. 2007).

9.1 What Is the MUS Approach?

Multiple-Use Water Services or MUS is an integrated and participatory approach in water services that considers (poor) consumers' actual water needs as a starting point in order to design, finance and manage existing or new water infrastructure for multiple domestic and productive purposes (Van Koppen et al. 2006; Renwick et al. 2007).

9.2 Why Apply MUS?

MUS is a relatively new approach, which has received a great deal of acceptance among policymakers, programme managers, investing organisations, water professionals and academic institutions (Adank et al. 2012). Multiple use water services can be more expensive compared to single-use services but have the potential to provide a wide range of economic and social benefits to consumers (Renwick et al. 2007):

- Increased income and multiple social benefits (improved human health, more and safe food, time savings, social equity) for more people;
- Vulnerability and poverty reduction;
- Improved sustainability of service delivery as MUS effectively target users' needs and priorities and result in increased income, encouraging communities to operate, sustain and finance services better.

Example: Multiple Sources for Multiple Water Uses in Urban Areas (WI 2012)

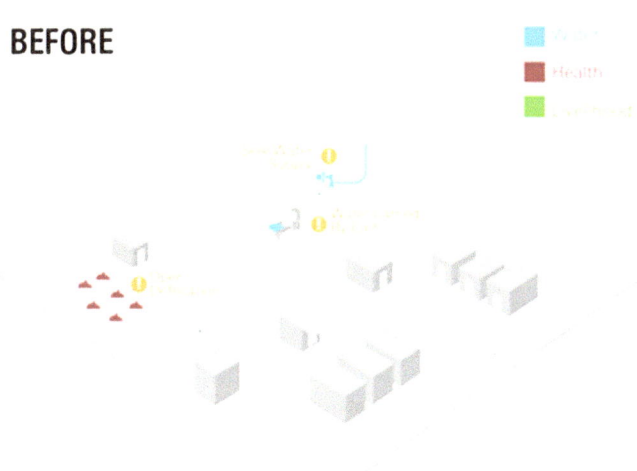

A single tap provides drinking water to consumers.

Water transport is distant leading to higher cost of water and often long waiting time.

Phenomena of diarrhoea are usual because of open defecation and limited hygiene.

Enterprises use less water due to high cost.

The installation of more public tap stands close to households increases available water and reduces wait time and cost for water transport.

With increased available water in addition to hygiene education and composting toilet application, health improves.

AFTER

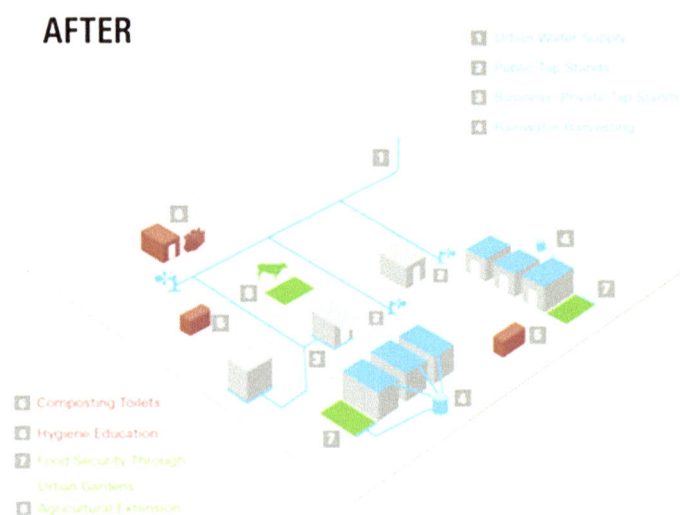

Businesses can also benefit from more, closer and cheaper water.

Rainwater harvesting schemes can provide water for household needs and commercial gardens.

Commercial gardening boosts income generation, reinforcing water infrastructure sustainability.

10 How Can We Calculate the Value of MUS Systems?

10.1 An Ecosystem Services Approach

Ecosystem Services (ESS) are the conditions and processes through which natural ecosystems and the species that make them up sustain and fulfil human life. Ecosystem services are also defined as all benefits people receive from ecosystems and can be used to describe connections between nature and human welfare (MA 2005). Ecosystem services changes result in outcomes, benefits or harms that people value, introducing the need of valuation and quantification of social welfare. Ecosystems and their functions and processes provide outputs of goods and services, which generate benefits to human populations that can then be measured as increases in human well-being (EFTEC 2005; Fig. 7).

Fig. 7 Classification of ecosystem services (*Source* TEEB (2010c))

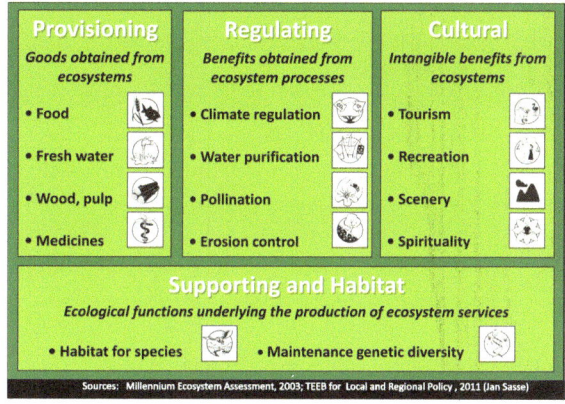

Fig. 8 Integration of ecosystem services into decision-making (*Source* Daily et al. (2009))

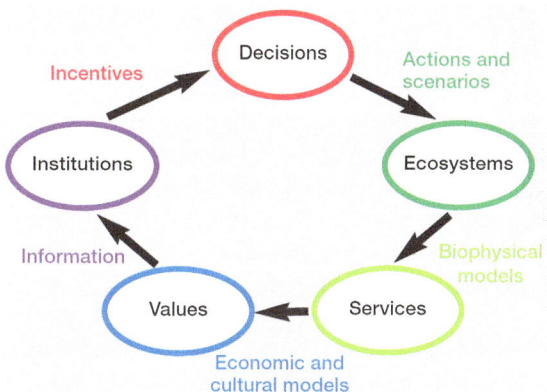

10.2 An Ecosystem Approach for Decision-Making

An ecosystems approach provides a framework for looking at whole ecosystems in decision-making and valuing the ecosystem services they provide, to ensure that society can maintain a healthy, resilient, natural environment now and in the future (Department for Environment 2013). The framework links the social, environmental and economic impact of an activity and evaluates it accordingly. It is a holistic approach that looks and quantifies not only costs but also benefits ranging from aesthetical pleasure and species preservation to job creation and property value. Carrying out economic valuation of the ecosystem services involved will help you to incorporate the value of the natural environment in your decision (Fig. 8).

The Challenge

Ascribing values to ecosystem services is not an end in itself, but rather one small step in the much larger and dynamic arena of political decision-making. Our challenge today is to build on this foundation and integrate ecosystem services into everyday decisions. This requires a new focus on services beyond provisioning services; an understanding of the interlinked production of services; a grasp of the decision-making processes of individual stakeholders; integration of research into institutional design and policy implementation; and the introduction of experimentally-based policy interventions designed for performance evaluation and improvement over time (Daily et al. 2009).

10.3 Economic Valuation of Ecosystem Services

Economic valuation is broadly accepted as an approach that can effectively link ecology and economics to evaluate benefits of management options. It can capture a broad array of environmental values, attributing not only a commercial value (e.g. the monetary value of timber) to an ecosystem service (NRC 2004), but also including many components that have no commercial or market basis (e.g. the aesthetic value of a natural landscape).

Given that, most ecosystem services are not sold in market, economic valuation techniques can be used to attach an appropriate value to their resulting benefits. Thus, economic valuation provides a systematic way in which environmental values can be factored into choices for better environmental decision-making. Other frameworks to valuate ESS are based on ecological or socio-cultural approaches. By 'values', we mean an attribute of a service or good, while valuation is the process of quantifying this attribute. The term 'economic value' describes the change in human wellbeing—welfare generated by a product.

10.4 How to Value Ecosystem Services?

To increase their total well-being people express preferences stemming from both use and non-use values. The sum total of use and non-use values related to a resource or an aspect of the environment is called Total Economic Value (TEV) and offers a useful framework to value ecosystem services. The metric for quantifying economic values is usually money (Fig. 9).

Use values encompass **direct use** values; consumptive (e.g. value of timber, fish etc.) or non-consumptive (e.g. recreation, aesthetics) and **indirect use** values that

Fig. 9 Components of total economic value

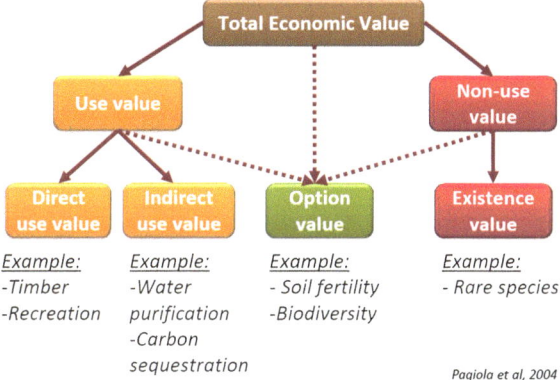

Pagiola et al, 2004

relate to the services provided by nature (e.g. air and water purification, erosion prevention) (de Groot et al. 2010). **Non-use values** is the importance attributed to an aspect of the environment in addition to or irrespective of its use values and could be described as the value attributed to its simple existence. **Option value** is when an individual derives benefit from ensuring that ecosystem services will be available for use in the future (EFTEC 2005).

Economic Valuation Techniques (de Groot et al. 2010)

A number of ways exist to translate economic and some socio-cultural values of ESS into monetary values. Market prices (marginal values) exist for many ecosystem services, especially the provisioning services such as timber and non-timber forest products. Values of other services are often also expressed through the market but in an indirect way, e.g. through (avoided) damage cost methods (for regulating services), hedonic pricing (influence of environmental attributes on property value) or travel cost methods for some cultural services such as aesthetically pleasing landscapes. Other alternatives are contingent valuation (e.g. questionnaires measuring preferences) and benefit transfer (i.e. using data from comparable studies).

TEV is a useful approach even if we cannot determine monetary values for all benefits. Having a monetary value for some benefit categories may be enough justification for choosing a conservation option over a more resource-exploitative alternative. In most cases, a partial monetization is more likely, more feasible and quite possibly less risky. By less risky, we mean that any analysis must be credible if stakeholders are to accept its findings (TEEB 2010a; Fig. 10).

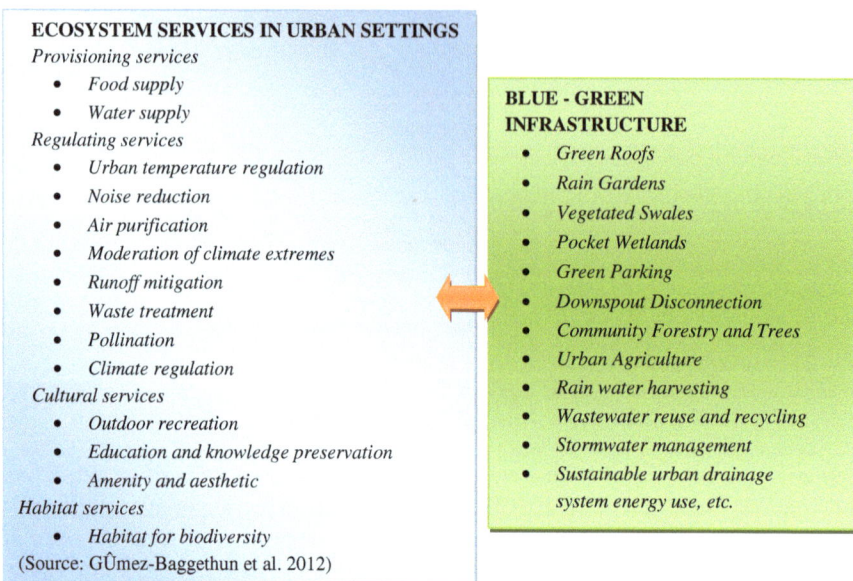

Fig. 10 Interactions between ecosystem services and blue-green infrastructure

Implement the Approach Step by Step (TEEB 2010b)

1. *Specify and agree on the problem*
   This is often a worthwhile effort because views can differ substantially. If key stakeholders share a common understanding of the problem, serious misunderstandings during the decision-making process and implementation can be avoided.
2. *Identify which ecosystem services are relevant*
   Ecosystem services are often interconnected. Identifying which ones are most important to your problem focuses the analysis. Going one by one through the list of services is a simple approach.
3. *Define the information needs and select appropriate methods*
   The better you can define your information needs beforehand, the easier it is to select the right analytical method and interpret the findings. Assessments differ in terms of which services are considered, the depth of detail required, timelines, spatial scope and monetization of the results and other factors. The study design determines what kind of information you get.
4. *Assess expected changes in availability and distribution of ecosystem services*
   If possible, use experts. Also, draw on fieldwork and documented experience from analyses in comparable settings. Use common sense and consult with colleagues on possible changes and their consequences, starting with the most obvious ecosystem services.
5. *Identify and appraise policy options*
   Based on the analysis of expected changes in ecosystem services, identify potential responses. Appraise these in terms of their legal and political feasibility

as well as their potential in reaching the targeted quality, quantity and combination of ecosystem services produced by your → *natural capital*.

6. *Assess distributional impacts of policy options*
   Changes in availability or distribution of ecosystem services affect people differently. This should be considered in social impact assessment, either as part of the analysis or as part of appraising policy options.
   Look out: The relative importance of each step is determined by your situation and objectives. Taken together, adapted to your needs, and incorporated into existing decision-making procedures, they offer guidance for considering natural capital in local policy. Other technical, legal, economic and social information also needs to be considered. The steps can also help you design a monitoring system and thereby track the condition of your natural capital.

10.5 How Can Economic Valuation Assist Policymaking?

- Providing information about benefits (in monetary terms or otherwise) and costs;
- Creating a common language for policymakers, business and society allowing the real value of ecosystems services to become visible and be accounted in decision making;
- Revealing the opportunities to work with nature by demonstrating where it offers a cost effective means of providing valuable services (e.g. water supply or reduced flood risk);
- Emphasising the urgency of action through demonstrating where and when the prevention of biodiversity loss is cheaper than restoration or replacement;
- Generating information about value for designing policy incentives (to reward the provision of ecosystem services and activities beneficial to the environment, to create markets or level the playing field in existing markets, and to ensure that polluters and resource users pay for their environmental impacts). (*Source* TEEB (2010c))

The Project (TEEB 2010b)

A change in national legislation has increased treatment requirements by lowering acceptable bacterial levels. The added designation of new residential areas will also increase volume to a level that can no longer be handled by your city's plant

Step 1   As director of the responsible department, you commission a pre-feasibility study for the construction of a modern plant that meets both quality and quantity requirements. The province-level development bank has an attractive credit scheme to help finance converting an agricultural site, but the costs are high and would require a considerable portion of the city's infrastructure budget. The city council agrees that an alternative solution is needed.

Step 2   At a workshop, you learn about the utility of wetlands for wastewater treatment. This helpful coincidence makes you realise what a preliminary ecosystem services appraisal would have shown: There is a wetland in your city close to an abandoned railroad track, which is neither accessible nor attractive.

Step 3   You invite the workshop expert who tells you that the location and condition of your wetland are suitable. He recommends you to determine how much rainwater run-off can be redirected to the wetland for rehabilitation, to examine flood control needs for neighbouring settlements and to establish whether redirected waters will reduce the volume flowing to the old plant.

Step 4   A team of colleagues consults available data for assessing the ecosystem services involved.

Step 5   Subsequent calculations reveal that this plan is considerably less costly than constructing a new treatment plant.

Step 6   It has the added benefit of liberating funds for other infrastructure projects and will not increase citizens' water bills. The area is uninhabited and unused, so an impact analysis on current users is unnecessary.

A local NGO agrees to help plant the reconstructed wetland and you convince the earthworks company to remove the railroad tracks to make space for a cycling and walking path.

Conclusions

The need to replace or construct new infrastructure presents an opportunity to examine ways to invest in more green, instead of grey, infrastructure or at least redesign projects in order to minimise damages to ecosystem services and biodiversity. There are many such opportunities in water provisioning (catchment management instead of water treatment plants), flood regulation (flood plains or mangroves rather than dykes) and landslide prevention (maintaining slopes covered with vegetation). Green infrastructure usually provides additional ecosystem services such as recreational value (habitat service)

11 Tools for Supporting Multiple-Use Water Services

MUS interventions provide an integrated urban water and urban green design, operation and management approach for sustainable cities. This more holistic approach would present a win-win scenario, in which urban green would be utilised as infrastructure for water services (e.g. mitigating urban floods) while urban water infrastructure would be used as irrigation source for urban green, increasing their performance in a range of services including amenities, reducing heat island effect and increasing ecosystem services. The urban water cycle is a complex system driven by time varying and stochastic inputs (rainfall, water demand). Thus, specialised models are required to support the optimal design, operation and management of urban water networks.

One of the most prominent urban water modelling tools that employs to some extent combined modelling of blue and green assets is **UVQ** (Mitchell and Diaper 2010). UVQ runs with daily time step to estimate the amount of water required for irrigating green areas and can estimate the reduction of potable water required for irrigation in case treated wastewater and/or harvested rainwater are used supplementary to potable.

Another model that can be used to study some urban water flows involved in MUS concept is **Aquacycle**, a daily urban water balance model developed to simulate the total urban water cycle as an integrated whole and investigate the potential use of locally-generated storm water and wastewater as a substitute for imported water. It can model from a single land block, such as a residential property, to an entire urban catchment (Mitchell 2005).

Music (Model for Urban Storm water Improvement Conceptualisation) is another MUS-related tool, specialised in helping urban storm water professionals visualise and compare possible strategies to tackle urban storm water hydrology and pollution impacts. Music allows the comparison of storm water management measures in order to achieve the best water quality, hydrology and cost outcomes. Music incorporates the recent findings of the Facility for Advancing Water Biofiltration (FAWB) to provide more accurate prediction of filtration-based treatment measures, especially bioretention and infiltration systems (MUSIC 2013).

The EPA **SWMM** is a storm water management model used for studying single events or continuous simulation of run-off quantity and quality from urban areas. The run-off component of SWMM operates on a collection of subcatchment areas that receive precipitation and generate run-off and pollutant loads. The routing portion of SWMM transports this run-off through a system of pipes, channels, storage/treatment devices, pumps and regulators. SWMM tracks the quantity and quality of run-off generated within each subcatchment, and the flow rate, flow depth and quality of water in each pipe and channel during a simulation period comprised of multiple time steps (Rossman 2010).

WASP is a decision☐supporting tool for developing policies and management protocols for sustainable irrigation of urban landscapes such as parks, sporting ovals, golf courses, etc. The name WASP comes from the acronym for Water☐

Atmosphere☐Soil☐Plant. WASP is designed to estimate monthly values of required irrigation specific to the environment (soil, macro☐climate & micro☐ climate characteristics), composition (planting characteristics) and function of the urban landscape (type of landscape outcome such as premium lush, moderate green or low maintenance), corresponding to different climatic years such as wet, dry and average years (IF Technologies 2013).

The previous models are versatile and very efficient for the type of applications they are intended for. However, these models do not offer the holistic approach required to explore the vision of blue-green services fully. For example, from all previous models, only UVQ can estimate both the irrigation needs and the portion of this demand that can be covered by storm water. However, UVQ does not offer a fine time step (time step fixed to 1 day) to simulate the peaks of run-off discharge. Furthermore, none of these models provides a metric to quantify the mitigation of the urban heat island effect.

UWOT is a bottom-up (micro-component based) urban water cycle model, which simulates demand at multiple time steps starting at the water appliance level. Most urban water models use a hydraulics-based conceptualisation of the urban water network, simulating actual water flows, including run-off, potable water and wastewater. UWOT uses an alternative approach based on the generation, aggregation and transmission of a demand signal, starting from the household water appliances and moving towards the source. The simulation results in the estimation of: (i) potable water demand, (ii) water level changes inside the tank and reservoirs, (iii) leakages, (iv) evaporation, (v) run-off, (vi) energy consumption (including both energy required for water circulation (e.g. pump of rainwater inside tank) and energy consumed by the water appliances (e.g. heat water for showering) and (vii) capital and operational costs. More details on UWOT can be found in the publications of Makropoulos et al. (2008), Rozos and Makropoulos (2012, 2013) and Rozos et al. (2013).

UWOT can be used in a wide range of urban water cycle applications representing any type of urban water network. Like any specialised model, a certain level of expertise is required to prepare a new UWOT project. To help beginners set up a new project, a simplified GUI was prepared serving as a front end to the UWOT engine, which runs seamlessly a set of predefined urban water networks (four predefined networks at household level and two at development level). An example of these predefined UWOT networks are shown in Fig. 11. This custom UWOT is called **MUS-Designer**.

MUS-Designer can simulate MUS technologies both at household and at development level. At household level, MUS-Designer simulates the potable water demand, the evaporative cooling (i.e. the energy absorbed from the environment during evapotranspiration) and the electric energy consumption of water appliances. The household water network can be conventional or include Best Available Technologies Not Entailing Excessive Costs (BATNEEC) and/or rainwater recycling and/or grey water recycling. The recycled water is used for toilet flushing and washing machines as well as for garden and green roof irrigation (Fig. 12).

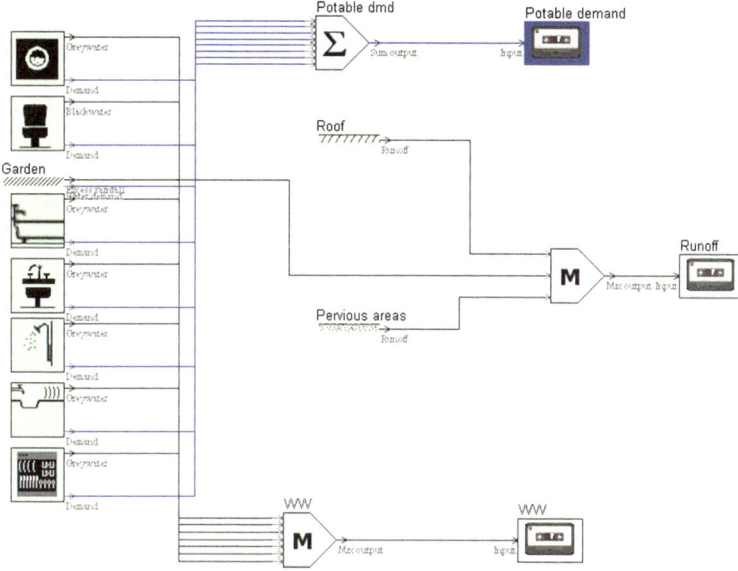

Fig. 11 Water network representation in UWOT of a conventional household

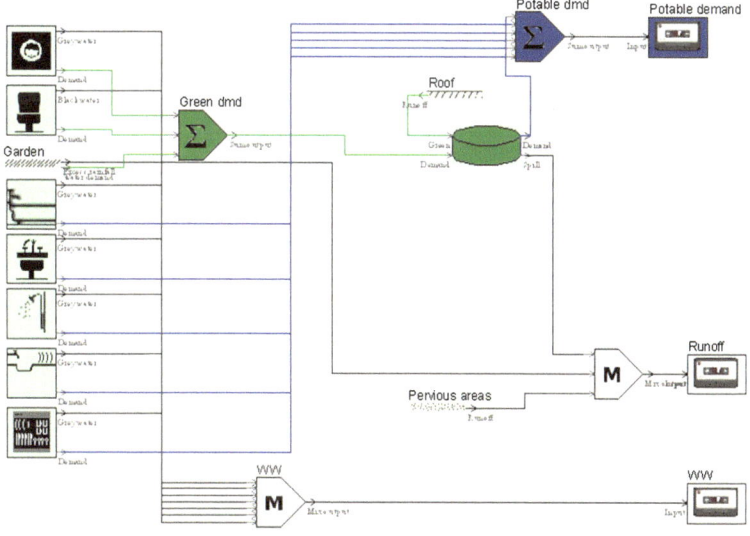

Fig. 12 Water network in UWOT of a household implementing rainwater recycling

At development level, MUS-Designer simulates the potable water demand, the water storage inside service reservoir (used for design purposes to estimate the required average annual inflow and reservoir capacity), the run-off from the development, the evaporative cooling and the solar energy on the development surface. The last two can be used to derive a metric regarding the heat island reduction benefit that green areas offer to urban environment.

The MUS-Designer, via a user-friendly interface, simulates the urban water cycle fluxes related to both blue services (potable demand, run-off volume) and green services (irrigation needs, run-off volume). Then, provides metrics related to the performance of these two services and to the benefits derived from their integrated operation (reduction of required raw water inflow, run-off mitigation and reduction of heat island effect).

Keywords and Definitions

ESS	Ecosystem Services
Green roof	Roof of a building that is entirely or to an extent covered with vegetation planted over a waterproofing membrane
Green walls	Vertical plants either grown on freestanding structures or attached to interior or exterior walls
Infiltration trench	Excavation filled with permeable material, such as rock and gravel, which is used to capture, treat, store and infiltrate storm water, enhancing the natural capacity of the ground to store and drain
MUS	Multiple use water services
MUSIC	Model for urban storm water improvement conceptualisation
Permeable pavement	Method of paving that allows water to infiltrate into the ground as it falls rather than running off into piped storm water drainage system
Rain gardens	Shallow planted depressions designed to receive rainwater from hard surfaces such as roofs, paved areas or roads
Retention ponds	Provide both storm water attenuation and treatment, while supporting emergent and submerged aquatic vegetation along their shoreline
Swales	Shallow, broad and vegetated channels filled with porous filter media to provide on-site treatment of storm water run-off
SWMM	Storm water management model
Urban agriculture	Practice of cultivating crops for food in cities
Urban water management	Emphasises decentralised storm water and rainwater management, such as Low Impact Development (LID) for USA, Decentralized Urban Design (DUD) for Germany, Water

	Sensitive Urban Design (WSUD) of Australia, Sound Water Cycle on National Planning (SWCNP) for Japan and Smart Watergy City (SWC) for South Korea
UWOT	Bottom up (micro-component based) urban water cycle model that uses an alternative approach based on the generation, aggregation and transmission of a demand signal, starting from the household water appliances and moving towards the source
WASP	Water-atmosphere-soil-plant

References

Adank M, van Koppen B, Smits S (2012) Guidelines for planning and providing multiple—use water services. IRC International Water and Sanitation Centre and International Water Management Institute. http://imawesa.info/wp-content/uploads/2012/11/MUS-Guidelines-for-Planning-and-Providing.pdf. Accessed 14 April 2014

Berghage RD, Beattie D, Jarrett AR, Thuring C, Razaei F, O'Connor TP (2009) Green roofs for stormwater runoff control. National Risk Management Research Laboratory Office of Research and Development, U.S. Environmental Protection Agency, Cincinnati. EPA/600/R-09/026. http://nepis.epa.gov/Adobe/PDF/P1003704.PDF. Accessed April 14, 2014

CGBC—Cascadia Green Building Council (2011) Toward net zero water: best management practices for decentralized sourcing and treatment. http://www.ecobuildingpulse.com/Images/TNZW_tcm131-1075029.PDF. Accessed 14 April 2014

Charles River Watershed Association (2008) Low impact best management practice (BMP) information sheet. http://www.crwa.org/projects/bmpfactsheets/crwa_raingarden.pdf. Accessed 14 April 2014

Clark C, Adriaens P, Talbot FB (2008a) Green roof valuation: a probabilistic economic analysis of environmental benefits. Environ Sci Technol 42(6):2155–2161

Clark M, Acomb G, Bean E (2008b) University of Florida permeable pavement fact sheet. http://buildgreen.ufl.edu/Fact_sheet_Permeable_Surfaces.pdf. Accessed 14 April 2014

Clear Water (2012a) Raingarden design principles. Quick reference guide. http://www.clearwater.asn.au/. Accessed 14 April 2014

Clear Water (2012b) Royal Park stormwater harvesting project city of Melbourne Royal Park, Melbourne, Victoria. http://www.clearwater.asn.au/user-data/case-studies/plans-designs/Royal-Park-Case-Study_FINAL.pdf. Accessed 14 April 2014

Continuing Education Centre (2014) Green walls: Integrating nature into buildings. http://continuingeducation.construction.com/crs.php?L=260&C=808. Accessed 14 April 2014

Daily CG et al (2009) Ecosystem services in decision making: time to deliver. Front Ecol Environ 7(1):21–28

de Groot RS et al (2010) Challenges in integrating the concept of ecosystem services and values in landscape planning, management and decision making. Ecol Complex 7:260–272

Department for Environment, Food & Rural Affairs (2013) Ecosystem services: guidance for policy and decision makers on using an ecosystems approach and valuing ecosystem services. www.gov.uk. Accessed 28 November 2013

DEPI—Department of Environment and Primary Industries (2014) Victoria's guide to green roofs, walls and facades, State of Victoria. http://imap.vic.gov.au/uploads/Growing%20Green%20Guide/Growing%20Green%20Guide%20FINAL%20DRAFT%20website4.pdf. Accessed April 14, 2014

EFTEC, Economics for the Environment Consultancy (2005) The economic, social and ecological value of ecosystem services: a literature review. The Department for Environment, Food and Rural Affairs (Defra), London

Environmental Protection Agency, Victoria (2008) Maintaining water sensitive urban design elements. http://www.epa.vic.gov.au/ ~ /media/Publications/1226.pdf. Accessed 14 April 2014

FAWB, Facility for Advancing Water Biofiltration (2008) Advancing the design of stormwater biofiltration.   http://www.monash.edu.au/fawb/products/fawb-advancing-rain-gardens-workshop-booklet.pdf. Accessed 20 December 2013

Fletcher T, Duncan H, Lloyd S, Poelsma P (2003) Stormwater flow and quality and the effectiveness of non-proprietary stormwater treatment measures. Draft report for the NSW EPA. April 2003. Cooperative Research Centre for Catchment Hydrology, Melbourne

GRHC—Green Roofs for Healthy Cities (2014) Official website. http://www.greenroofs.org/. Accessed 14 April 2014

Herman R (2003) Green roofs in Germany: yesterday, today and tomorrow. In Proceedings of 1st North American green roof conference: greening rooftops for sustainable communities, Chicago. May 29–30, 2003. The Cardinal Group, Toronto, pp. 41–45

Hobart City Council (2006) Water sensitive urban design site development guidelines and practice notes.   Hobart.   http://www.hobartcity.com.au/content/InternetWebsite/Environment/Stormwater_and_Waterways/Water_Sensitive_Urban_Design.aspx. Accessed 14 April 2014

IF Technologies (2013) About WASP online tool. Cooperative Research Centre for Irrigation Futures

IGRA (2012) Green roofs news. International Green Roof Association Global Networking for Green   Roofs.   http://www.igra-world.com/links_and_downloads/images_dynamic/IGRA_Green_Roof_News_2_2012.pdf. Accessed 14 April 2014

Köhler M (2004) Green roof technology—from a fire-protection system to a central instrument in sustainable urban design. Second green roof conference, Portland, Oregon. http://www.greenroofs.org/grtok/sbp_browse.php?id=20&what=view. Accessed 14 April 2014

Leinster S (2004) Shaun Leinster, Senior Engineer, Ecological Engineering Pty Ltd, Brisbane, Queensland, personal communication

Liu K, Baskaran B (2003) Thermal performance of green roofs through field evaluation. National Research Council, Institute for Research in Construction, Ottawa. http://archive.nrccnrc.gc.ca/obj/irc/doc/pubs/nrcc46412/nrcc46412.pdf. Accessed 14 April 2014

Lloyd SD, Wong THF, Chesterfield CJ (2002) Water sensitive urban design—a stormwater management perspective. Industry report no 02/10. Cooperative Research Centre for Freshwater Ecology, Melbourne

MA, Millennium Ecosystem Assessment (2005) Millennium ecosystem assessment—ecosystems and human well-being: synthesis. Island Press, Washington, DC

Maimon A, Tal A, Friedler E, Gross A (2010) Safe on-site reuse of greywater for irrigation—a critical review of current guidelines. Environ Sci Technol 44:3213–3220

Makropoulos CK, Natsis K, Liu S, Mittas K, Butler D (2008) Decision support for sustainable option selection in integrated urban water management. Environ Model Softw 23(12):1448–1460

McPherson EG, Simpson JR, Peper PJ, Maco SE, Xiao Q (2005) Municipal forest benefits and costs in five US cities. J Forest 103(8):411–416

Mitchell VG (2005) Aquacycle: a daily urban water balance model. Monash University CRC for catchment hydrology

Mitchell VG, Diaper C (2010) UVQ user manual. CMIT report no 2005-282, CSIRO, June

MUSIC (2013) MUSIC (model for urban stormwater improvement conceptualisation). http://www.waterforliveability.org.au/?page_id=2531. Accessed 14 April 2014

Nolasco J (2011) Sustainable water management for urban agriculture: planting justice. Pacific Institute,   Oakland.   http://libertyconcepts.com/wp-content/uploads/sites/21/2013/02/sustainable_water_management_for_urban_agriculture3.pdf. Accessed 14 April 2014

NRC, National Research Council, Committee on Assessing and Valuing the Services of Aquatic and Related Terrestrial Ecosystems (2004) The meaning of value and use of economic valuation in the environmental policy decision-making process. Valuing ecosystem services: toward better environmental decision-making. National Academy of Sciences

Peck SW, Kuhn M (2003) Design guidelines for green roofs. Ontario Association of Architects, Canada. http://www.cmhc-schl.gc.ca/en/inpr/bude/himu/coedar/upload/Design-Guidelines-for-Green-Roofs.pdf. Accessed 14 April 2014

Peck SW, Callaghan C, Kuhn, ME, Bass B (1999) Greenbacks from green roofs: forging a new industry in Canada. Prepared for the Canada Mortgage and Housing Corporation, Ottawa. http://carmelacanzonieri.com/3740/readings/greenroofs%2Bgreen%20design/Greenbacks%20from%20greenroofs.pdf. Accessed 14 April 2014

Pennsylvania Department of Environmental Protection (2006) Stormwater BMP manual. http://www.elibrary.dep.state.pa.us/dsweb/Get/Document-68851/363-0300-002.pdf. Accessed 14 April 2014

Perini K, Rosasco P (2013) Cost-benefit analysis for green façades and living wall systems. Build Environ 70:110–121. doi:10.1016/j.buildenv.2013.08.012. http://scbrims.files.wordpress.com/2013/10/111013-cost-benefit-analysis-for-green-facades-and-lws.pdf. Accessed 14 April 2014

Prince George's County, Maryland (2014) Guidelines for permeable pavement. http://www.princegeorgescountymd.gov/sites/StormwaterManagement/Resources/BMP/Documents/4_Guidelines%20for%20Permeable%20Pavement.pdf. Accessed 14 April 2014

Renwick M et al (2007) Multiple use water services for the poor: assessing the state of knowledge. Winrock International, Arlington. http://www.winrockwater.org/docs/Final%20Report%20Multiple%20Use%20Water%20Services%20Final%20report%20feb%2008.pdf. Accessed 14 April 2014

Rossman LA (2010) Storm water management model user's manual version 5.0. Water Supply and Water Resources Division, National Risk Management Research Laboratory, United States Environmental Protection Agency, Cincinnati

Rozos E, Makropoulos C (2012) Assessing the combined benefits of water recycling technologies by modelling the total urban water cycle. Urban Water J 9(1). doi:10.1080/1573062X.2011.630096

Rozos E, Makropoulos C (2013) Source to tap urban water cycle modelling. Environ Modell Softw 41:139–150. doi:10.1016/j.envsoft.2012.11.015, Elsevier

Rozos E, Makropoulos C, Maksimović Č (2013) Water science & technology: water supply (in press). Rethinking urban areas: an example of an integrated blue-green approach 13(6):1534–1542. doi:10.2166/ws.2013.140

TEEB, The Economics of Ecosystems and Biodiversity (2010a) The economics of ecosystems and biodiversity for local and regional policy makers. Progress Press, Malta

TEEB, The Economics of Ecosystems and Biodiversity (2010b) The economics of ecosystems and biodiversity for local and regional policy makers. Progress Press, Malta

TEEB, The Economics of Ecosystems and Biodiversity (2010c) Mainstreaming the economics of nature: a synthesis of the approach, conclusions and recommendations of TEEB. Progress Press, Malta

Tolderlund L (2010) Design guidelines and maintenance manual for green roofs in the semi-arid and arid west. http://www2.epa.gov/sites/production/files/documents/GreenRoofsSemiAridAridWest.pdf. Accessed 14 April 2014

UN ESCAP (2012) Low carbon green growth roadmap for Asia and the Pacific, Background policy paper. Water resource management: policy recommendations for the development of eco-efficient infrastructure. Prepared by: Dr Reeho Kim, Dr Jungsoo Mun and Dr Jongbin Park, Korea Institute of Construction Technology. United Nations publication 2012

University of Maryland Extension (2011) Permeable pavement fact sheet information for Howard County, Maryland homeowners. https://extension.umd.edu/sites/default/files/_docs/programs/master-gardeners/Howardcounty/Baywise/PermeablePavingHowardCountyMasterGardeners10_5_11%20Final.pdf. Accessed 14 April 2014

URS (2003) Water sensitive urban design technical guidelines for Western Sydney. Draft report prepared for the Upper Parramatta River Catchment Trust, Sydney, New South Wales

US EPA (2008) Clean watersheds needs survey 2008—report to congress. United States Environmental Agency. EPA-832-R-10-002

US EPA (2013). Rainwater harvesting: conservation, credit, codes, and cost literature review and case studies. United States Environmental Protection Agency, EPA-841-R-13-002. http://water.epa.gov/polwaste/nps/upload/rainharvesting.pdf

US EPA (2014a) Green roofs. Environmental Protection Agency, US. http://www.epa.gov/heatisland/mitigation/greenroofs.htm#3. Accessed 14 April 2014

US EPA (2014b) Vegetated filter strip. Environmental Protection Agency, US. http://cfpub.epa.gov/npdes/stormwater/menuofbmps/index.cfm?action=factsheet_results&view=specific&bmp=76. Accessed 14 April 2014

Van Koppen B, Moriarty P, Boelee E (2006) Multiple-use water services to advance the millennium development goals. Research Report 98. International Water Management Institute, Colombo. http://www.musgroup.net/page/1056. Accessed 14 April 2014

WERF—Water Environment Research Foundation (2009) BMP and LID whole life cost models, version 2.0. Alexandria, VA. https://www.werf.org/i/a/Ka/Search/ResearchProfile.aspx?ReportId=SW2R08. Accessed 14 April 2014

WI, Winrock International (2012) A guide to multiple-use water services. http://www.rockefellerfoundation.org/uploads/files/6017a66b-db64-46ca-97ff-2db8e873cc04.pdf. Accessed 14 April 2014

Yu ZLT, Rahardianto A, DeShazo JR, Stenstrom MK, Cohen Y (2013) Critical review: regulatory incentives and impediments for onsite graywater reuse in the United States. Water Environ Res 85(7):650–662

Chapter 3
Case Studies Illustrating the Multiple-Use Water Services Options

Abstract This brief comprises case studies that illustrate blue-green options in urban locations in Seoul, Melbourne, Philadelphia, Wallington and Brisbane as well as projects conducted by nonprofit organisations in both Asia and Africa. Each case study breaks out the challenges, policy framework, benefits, benchmarks, lessons learned (success and failures) and potential next steps. The section on MUS, an integrated, participatory service delivery approach that considers water users' multiple needs as a starting point and involves design, financing, provision and management of sustainable water services along with supporting programmes regarding human health, livelihoods and environment, provides insight into programmes conducted in seven countries.

Keyword BedZED · DRMS · Green City, Clean Waters' plan · Green Stormwater Infrastructure (GSI) · Land-Water-Infrastructure · MUS · RHM · RWH · WSUD

1 'Four Alls for All': Policy Act on Decentralised Water Supply Through Rainwater Harvesting and Management Systems in Seoul

'Four alls for all' can be summarised as 'RWH for (and by) all people, collecting all available rainwater, by all possible methods, and in all possible locations, with the result that people and nature all win' (International Water Association (IWA) 2008), and reflects the newly established municipal policy towards RWH in Seoul. Inspired by the Star City project, this ordinance requires rainwater harvesting to be applied in any new or redevelopment of residential and commercial property, with the vision of transforming Seoul 'from a drain to a rain city' (Han 2012).

© The Author(s) 2015

Č. Maksimović et al., *Rethinking Infrastructure Design for Multi-Use Water Services*, SpringerBriefs in Environmental Science, DOI 10.1007/978-3-319-06275-4_3

1.1 What Is the Challenge?

The complete dependence of Seoul's water supply and drainage on a centralised system is unstable in view of climate change, increased urbanisation rates and the energy crisis. Water supply reliability is further challenged by the ageing infrastructure and increasing energy costs. Urban flooding risk has increased since the uptake of open spaces has increased run-off from 9 % (in pre-development) to 47 % and decreased infiltration from 40 to 23 %. Urban heat island effects are observed. Moving towards decentralised water management and developing such systems can reduce the prevailing risks.

1.2 What Is the Policy Framework?

The first law to promote Rainwater Harvesting and Management (RHM) in Korea is the 'Water Law' (2001), which forces sport facilities with roof area larger than 2,500 m^2 to have rainwater harvesting systems (RHS) (Han and Park 2005). Four world cup stadiums in Korea have been accordingly equipped under this Law, while numerous RHM projects have emerged. In 2010, the 'Law to promote and support water reuse' released by the Ministry of Environment, puts into effect the installation of RHS for non-potable uses. The policy development by municipalities is much more active than that by the central government in Korea. While no ordinate for RHM existed prior to 2003, 26 ordinates were adopted from 2004 to 2008, 19 more from 2008 to 2009 and as of July 2010, 47 ordinances exist for RHM by the municipalities. These vary in scope, the early ones focusing on rainwater storage and use, progressively expanding their focus to flood control (e.g. Seoul, Daejeon, Incheon cities) and further to integrated water management for restoring the hydrologic cycle (e.g. Suwon city) (UN ESCAP 2012). Various types of incentives have been promoted in parallel to secure the uptake of the ordinances: Seoul city financially supports the installation of RHS (up to 10 million won), Jeju island provides subsidies covering up to 80 % of total installation costs, some municipalities lower the costs of tap water supply and wastewater treatments for the people who use RHS, others reward a higher floor space index allowance to buildings with RHS (UN ESCAP 2012).

The paradigm of RHM is continuously evolving, currently supporting the multi-benefit integrated rainwater management for urban flood mitigation, water conservation, non-point source pollution control, hydrologic cycle restoration and urban heat island phenomena alleviation (UN ESCAP 2012).

1.3 What Is the Way Forward?

Decentralised Rainwater Management Systems (DRMS) can result in water conservation and demand reduction, mitigate the risk of flooding and reduce energy consumption. They act as a short-term and long-term solution for climate change adaptation, supplementing the existing urban water infrastructures. The scale of implementation of such systems varies from small to large and can be customised according to the site-specific needs.

As of July 2010, there are 659 RHMS in Korea, which are currently being operated or planned. Most of them are located near Seoul Metropolitan Area, including 520 facilities in Seoul city and 48 facilities in Gyeonggi-Do. Nearly 30 % of the total RHMS are installed in housing complexes and 21 % were installed in schools, while 26 % of them were constructed only for stormwater run-off control (UN ESCAP 2012). The collected rainwater is used for cleaning, irrigation, firefighting and toilet flushing, with different usage patterns across types and sites. In the housing complexes, rainwater stored is almost equally distributed for firefighting (36 %), irrigation and gardening (34 %) and cleaning (30 %) purposes. In schools, gardening is the main use (50 %), followed by cleaning (19 %) and toilet flushing (16 %), while smaller amounts are allocated to spraying of playgrounds (11 %) and firefighting (4 %) (UN ESCAP 2012).

1.4 What Are the Benefits of Implementing DRMS?

DRMS use natural water resources to render multiple benefits in the case of Korea and Seoul in particular, including: resolve water shortage problems, reduce damage from floods, improve river water quality by reducing non-point source pollutants, preserve groundwater levels, mitigate urban heat island effects, save energy and reduce emissions, contribute to firefighting (from the existing scattered rainwater tanks), create an eco-friendly environment and improve the quality of life for citizens by creating better surroundings and environment. They can support the city's vision towards an eco-friendly Seoul where man, nature and city co-exist, recovering water environment to pre-urbanisation level.

1.5 Benchmarking Example: The Star City Rainwater Harvesting and Management System

The Star City RHM System (Fig. 1), located in Gwangjin-gu (eastern section of Seoul), has been in operation since 2007 (construction started in October 2003) and is gaining worldwide attention as a model for futuristic water management and climate change adaptation system, which supplements the existing centralised water

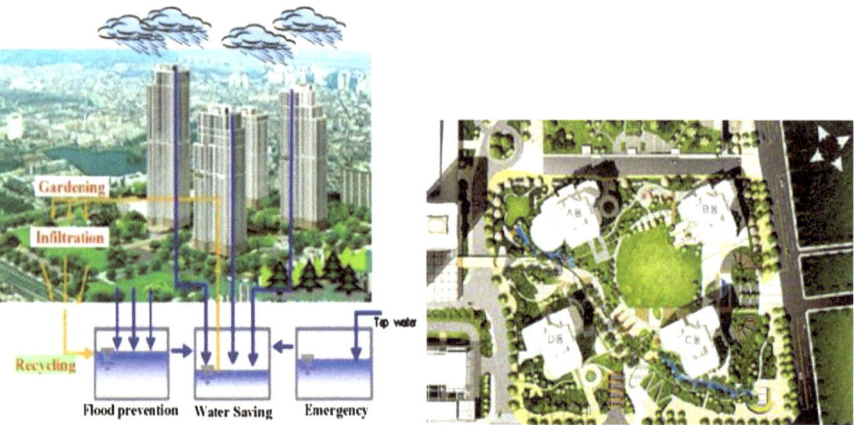

Fig. 1 Site plan (*left*) and aerial image (*right*) of the Star City in Seoul (*Source* UNEP and SEI 2009)

infrastructure (IWA 2008). Built on a 6.25 ha site (originally a playing field) owned by Konkuk University, Star City is a major real estate development project with more than 1,300 apartment units (accommodating 4,000–5,000 people) spread between four apartment blocks (each between 35 and 57 storeys) plus a department store. Each apartment block has a plan area of 1,500 m². The rainwater harvesting system has been designed by the Rainwater Research Center at Seoul National University and Professor Mooyoung Han. The catchment area consists of 6,200 square metres of rooftop and 45,000 square metres of terrace. Rainwater is piped to the RWH basement under the 35-storey block where there are three separate storage tanks each of 1,000 m³ capacity (3,000 m³ in total) (Fig. 1). The basic design idea of the RWH system was to collect up to the first 100 mm of rainwater that falls on the complex and to use the collected rainwater for gardening and public toilets. The first two tanks are used to collect rainwater from the rooftop and the ground and mitigate flood risk in the area during the monsoon season. A special feature is that most of the irrigated water in the garden is infiltrated into the ground and returns to the tank for multiple use. The third tank is used to store tap water in case of emergency.

The water conservation is expected to be approximately 40,000 m³ per year, which is about 67 % of the annual amount of rainfall over the Star City complex. The ratios of the volume of water conserved per month to the monthly amount of rainfall, ranged widely from 18 % in July to more than 200 % in November. More than 200 % could be achieved by supplying stored rainwater left from the previous month and by irrigation, which is recycled to the rainwater tank (Han and Mun 2011). The tank volume–catchment area ratio of 5.8 m³/100 m² and 10-year design period for this building would provide protection from a 50-year rainfall flood event. The risk of floods is controlled pro-actively with the remote control system, by emptying or filling the tanks appropriately. With regard to energy saving, approximately 8.9 MWh of electricity is expected to be saved per year by rainwater use in the Star City project.

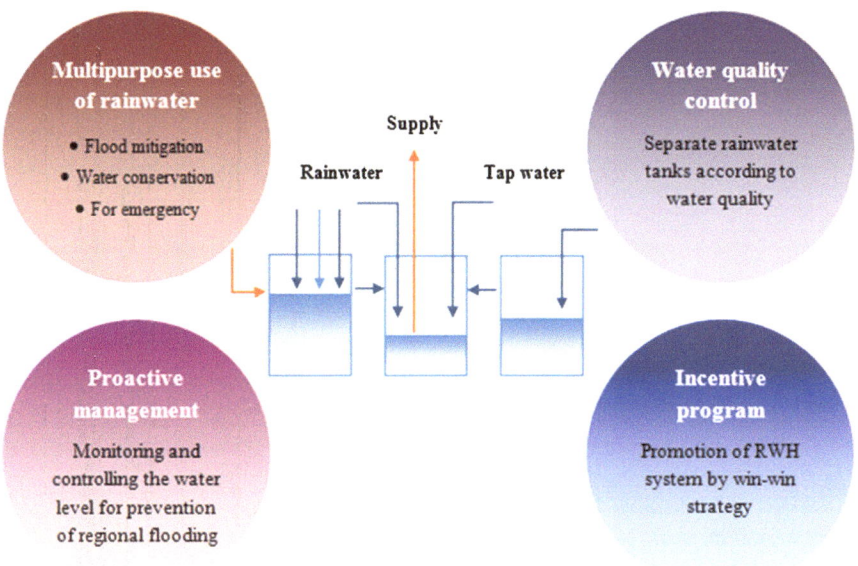

Fig. 2 Innovative elements of the Star City (*Source* Han 2008a)

The Star City project has four innovative pillars: multipurpose use of harvested rainwater, water quality control, proactive management and incentive programme. The social impact of the Star City was great, enacting regulations and the formulation of a citywide ordinance to promote more RWHS installations in development projects (Fig. 2).

1.6 Additional Demonstration Projects in Seoul

Since the establishment of the Korean Rainwater Catchment Systems Association in 2001, many successful demonstration projects have been implemented. Selected examples are presented in the following sections.

Raemian Apartment Complex in Seocho-gu, Seoul

Constructed: 2008, Facility Capacity: 2,000 m^3, Usage: Landscape (stream, water fall, jet of water), cleaning. (*Source* Han 2008b).

Xi Apartment Complex in Seocho-gu, Seoul

Constructed: 2006, Facility Capacity: 80 m^3, Usage: Landscape, cleaning (*Source* Han 2008b).

Seoul Village of Traditional Houses, Seoul

Storage capacity: 7,000 m^3
Before: Flood prevention (Store rainwater temporarily before discharging)
After: Transformed into multipurpose facility → (Rainy Season) Flood prevention + (Normal Days) Landscape water (Source: Han 2008b)

1.7 Which Are the Lessons Learned and Success Factors in the Case of Seoul?

- Inspiring examples such as the Star City RWHS can stimulate the regional policy process. 'A successful rainwater harvesting (RWH) project in the Korean capital, Seoul, has prompted a revision of water policy for the city as a whole and looks set to provide the impetus for increased interest in RWH in other Korean municipalities' (IWA 2008).
- Korea's new view on Rainwater Management: moving away from the collection and drainage focus, to maximised storage, infiltration and utilisation focused measures in the area.
- Promote the introduction of as many types of rainwater management facilities as possible, and as many sectors as possible (roads, lots, houses, parks, etc.). 'For all—Four all' Campaign: with all rainwater, through all facilities, in all sectors, with all participation.

- Participation from public and private sectors, a peaceful water management solution where all stakeholders win.
- Drafting of Strategic Plans for both the Public and Private sector:
  - Public (implementation from the city, local districts, investment organisations, introduce it from the planning/installing stage).
  - Private Sector (implementation in new town redevelopment project and land development projects, provide conditions when approving the projects).
- Provide benefits and incentives to the private sector to boost implementation: increase city subsidy in installing small-scale rainwater using facility, ease construction floor-area ratio: as compared to the basic floor-area ratio (max 5 %), grant exemption from local tax (max 20 %) to the owner of green building.
- Encourage participation by publishing rainwater guidelines, brochures, pamphlets, etc., engage in active communication with the public, develop massive citizen campaigns with civic groups.

2 Water Sensitive Urban Design in Lynbrook Estate, Melbourne, Australia

Lynbrook estate project is a green field (new-built) residential development incorporating Water Sensitive Urban Design principles (WSUD) at streetscape and sub-catchment scale. Lynbrook Estate lies 35 km southeast of Melbourne and is the first broad scale WSUD implementation in Melbourne, applying a 'treatment train', a sequence of stormwater treatment techniques, to manage stormwater (Farrelly and Davis 2009). It was established as a pilot project by Melbourne Water to test and demonstrate the applicability and effectiveness of WSUD regarding stormwater treatment.

2.1 What Was Implemented in Lynbrook?

The Lynbrook demonstration project incorporating WSUD stormwater treatment practices began construction in 1999 and covered an area of approximately 32 ha including 270 allotments and public open space. The treatment train constructed mainly comprised grass swales and underlying gravel trench that treat roof and street run-off and convey it to an artificial wetland before its final discharge into an ornamental lake. More specifically:

- Stormwater run-off from roofs and the local streets is captured, filtered and primarily treated through a combined system of grass swales and gravel trench

Fig. 3 Biofiltration systems (*left*) and artificial wetland (*right*) in Lynbrook Estate (*Source* wsud. org; Brown and Clarke 2007)

and is led to the main entrance boulevard though a perforated PVC pipe located at the base of the trench.

- The same biofiltration system (grass swales—gravel trench—pipe) is constructed through the median strip of Lynbrook Boulevard (Fig. 3), where the catchment run-off is also collected, filtered and treated.
- Treated stormwater inside the biofiltration system of the Boulevard is conveyed to a constructed wetland (Fig. 3), where it is secondary treated prior to final discharge in the ornamental lake.

2.2 What Were the Challenges?

The key challenges and impediments encountered were mainly due to the innovative nature of the project compared to the conventional and tested stormwater management practices in developing green field areas (Farrelly and Davis 2009). The main challenges involved:

- Lack of knowledge and familiarity with new unconventional sustainable technologies and related scepticism within the local government hampering the progress of the project.
- Raised concerns related to the overall cost of the proposed new technologies mainly regarding a possible failure.
- Limited evidence/data of similar tested experiences to smooth concerns and inspire confidence.

- Construction site problems due to lack of experience and awareness on new technologies of the people involved in the construction process.

2.3 Was the Project Successful? In What Way?

Overall, Lynbrook Estate is considered a successful demonstration project for the application of an innovative stormwater treatment system in Melbourne and one of Australia's WSUD leading examples (Farrelly and Davis 2009). The success of the project, mainly regarding collaboration across all involving partners as well as the positive commercial and technical performance results, affected significantly the improvement of sectoral confidence in Urban Stormwater Quality Management (Brown and Clarke 2007).

2.3.1 Facts

- Monitoring results indicated that the unconventional stormwater treatment systems applied in Lynbrook Estate present better hydraulic performance when compared to traditional drainage systems (Brown and Clarke 2007) and moreover, regarding water quality issues, provide a 60 % reduction of total nitrogen, 80 % reduction in total phosphorous and 90 % reduction of total suspended solids.
- Despite the earlier concerns of the developers on the consumers' reactions to the new technologies applied, sales were significantly increased shortly after the construction began, while sale prices of the subdivisions incorporating WSUD features reported to be increased by a rate of 20–30 %.
- A cost comparison study indicated that implementing WSUD stormwater drainage systems could be only 5 % more expensive while comparing to conventional systems (Lloyd 2001; Wong 2001).
- The local community considered the developments to be more 'aesthetically attractive' (Lloyd 2004) than the traditional ones.
- In 2000, Lynbrook won the President's Award in the Urban Development Institute of Australia Awards for Excellence and in 2001 the Cooperative Research Centres' Association Technology Transfer Award.

2.4 What Can We Learn from the Case of Lynbrook Estate?

- Demonstration of an effective combination of different stormwater treatment techniques.
- Importance of raising awareness and building knowledge among local authorities to address capacity deficits in urban water practices.

Fig. 4 The vision of Philadelphia (*Source* PWD, http://phillywatersheds.org)

- Significance of collecting and providing 'tangible' evidence on technical applicability and effectiveness, but also on socio-economic data, to secure acceptance and build confidence among community and local authorities when promoting Water Sensitive Urban Design practices or any other new technology.

3 Green City, Clean Waters: The Vision of Philadelphia

Differentiating from the traditional approach of collecting stormwater from the impervious surfaces of urban areas and quickly leading them away from the city to streams and rivers using huge and expensive 'grey' infrastructure, such as pipes, pumps and storage tanks, Philadelphia has a completely new vision for managing stormwater (Fig. 4).

Green City, Clean Waters Vision
To unite the City of Philadelphia with its water environment, creating a green legacy for future generations while incorporating a balance between ecology, economics and equity.

Through the 'Green City, Clean Waters' 25-year plan (from 2011 to 2035), the city of Philadelphia aims at the protection and quality upgrade of its watersheds applying wide-scale Green Stormwater Infrastructure.

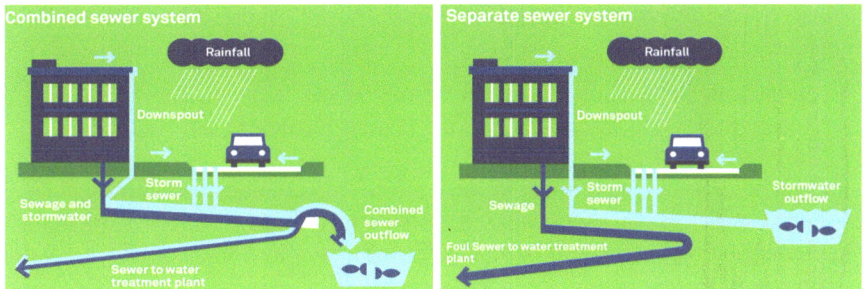

Fig. 5 Combined and separated sewer systems

3.1 What Is the Case?

In Philadelphia, the largest city in Pennsylvania, wastewater and stormwater is transferred to the city's Water Pollution Control Plants (WPCP) for treatment prior to final disposal to the nearby waterways via either Combined Sewer Systems (CSS) or Separated Sewer Systems (PWD 2013) as displayed in Fig. 5. All sewer systems as well as WPCP are owned and managed by the Philadelphia Water Department (PWD). The CSS, transporting both wastewater and stormwater through the same pipe, cover approximately 48 % of the total city area. During severe rainfall events, the flow may exceed CSS capacity and the overflow, known as Combined Sewer Overflow (CSO), is then diverted straight to the rivers or streams without treatment, causing pollution and threatening the aquatic environment.

3.2 What Was the Way Forward?

The 'Green City, Clean Waters' plan adopted by the city of Philadelphia is a 25-year plan aimed at reducing pollution from CSO to the extent that 85 % of the water would be treated in the CSS. The main goals of the plan are summarised below (PWD 2013):

- Enhancement of river quality, aesthetics and recreation.
- Restoration and improvement of aquatic habitats.
- Water quality improvement and water quantity decrease flows into the CSS.

To accomplish these targets, Philadelphia has developed and adopted an integrated 'Land-Water-Infrastructure' approach. This approach includes traditional 'Grey' Stormwater Infrastructure (i.e. sewers, pipes and treatment plants) as well as Green Infrastructure and projects for natural restoration of aquatic habitats. Green Stormwater Infrastructure (GSI) applies nature-complying methods to manage runoff at the source through soil and vegetation absorption and filtration of water. The

'Land-Water-Infrastructure' approach includes the following practices to be implemented:

- Large-scale application of GSI techniques in public areas.
- Requirements and incentives for GSI implementation in private lands.
- Large-scale tree planting.
- Increased access and enhanced recreational opportunities on waterways.
- Open spaces maintenance and use of them to manage stormwater on-site.
- Redevelopment of empty and abandoned areas and transformation to open spaces.
- Restoration of streams for the improvement of aquatic habitats.
- Grey stormwater infrastructure maintenance and improvement projects.

3.3 What Tools and Techniques for Green Stormwater Infrastructure Have Been Used?

Applying Green Stormwater Infrastructure on an intensively urbanised area such as Philadelphia necessitates an imaginative and decentralised planning and design approach incorporating a mixture of different tools. Such tools include reduction of impervious surfaces, infiltration and subsurface storage, green roofs, swales, tree planting, permeable/porous pavement, stormwater bump-outs and planter boxes, and have been widely used in Philadelphia (PWD 2013). Some of these tools are displayed in Figs. 6, 7, 8 and 9.

Most of the financing of the 25-year programme will be provided by the PWD, estimated at $2.4 billion, while private investments are expected to raise the total budget to $3 billion (PWD 2013). Moreover, incentives such as Parcel-Based Billing Initiative are planned to encourage the private sector to implement green stormwater infrastructure practices. The Parcel-Based Billing Initiative is a stormwater fee charged to non-residential properties depending on their impervious percentage coverage.

Finally, a cost–benefit analysis performed by the PWD to evaluate the best approach for CSO demonstrated that for equal investment levels and similar CSO volume reduction, the benefits resulted by the application of distributed Green Stormwater Infrastructure practices for a 40-year period, translated into economic value, would be 20 times more than that of conventional stormwater infrastructure such as large pipes and pumping stations. The estimated benefits included increased recreational opportunities, enhanced aesthetics and property value, air quality improvements, water quality and ecosystem improvement, green job creation and reduced urban heat stress.

For such an ambitious project to succeed and achieve its objectives, a collaborative procedure and the participation of numerous partners is required, including PWD, businesses, interest groups, citizens, civic associations and neighbourhood groups.

Fig. 6 Porous pavement (*left*) and tree trench (*right*) [*Source* CDM Smith (the support of CDM SMITH in providing reference material is kindly acknowledged)]

Fig. 7 Flow-through planter (*left*) and rain barrel (*right*) (*Source* PWD 2013)

Fig. 8 Swale and Rain Garden in Womrath Park (*Source* CDM Smith 2013)

3.4 What Are the Benefits Foreseen Following These Multiple-Use Water Services Strategies?

The expected outcomes target simultaneously environmental, social and economic (Triple Bottom Line approach) benefits:

- Protection and improvement of waterways and aquatic ecosystems.

- Air quality upgrade and lower temperatures in summer through greening of the city.
- Cooling effect of trees related to reduced energy.
- Reduced localised flooding by on-site management of stormwater.
- Reduction in health problems and deaths from the shift to a healthier urban environment.
- Increase in recreational activities and tourism.
- Enhanced aesthetics, social pride and quality of life.
- Job creation—around 250 new green jobs per year.
- Higher property values—residential property values near parks are expected to increase by $350 million. This would lead to higher tax revenues that can be returned to finance the programme.

3.5 What Can We Learn from Philadelphia?

- Demonstration of an innovative, cost-effective, blue-green solution to the problem of sewer overflows aimed at multiple environmental, economic and social benefits.
- Significance of city-wide planning frameworks.
- Concurrent policy efforts, appropriate regulations and incentives are significant tools to foster public awareness and stimulate wide-scale implementations.
- Community commitment and shared agendas across different stakeholders.

4 Integrated Water Recycling in Brisbane, Australia

Location: The Payne Road development is located in Brisbane's western suburbs. The development area is situated on the urban fringe and is bordered on the west by the Brisbane State Forest and Enoggera Reservoir Reserve, and on the northeast by the Enoggera Creek.

Fig. 9 Stormwater bump-out (*left*) and stormwater planter (*right*) (*Source* PWD 2013)

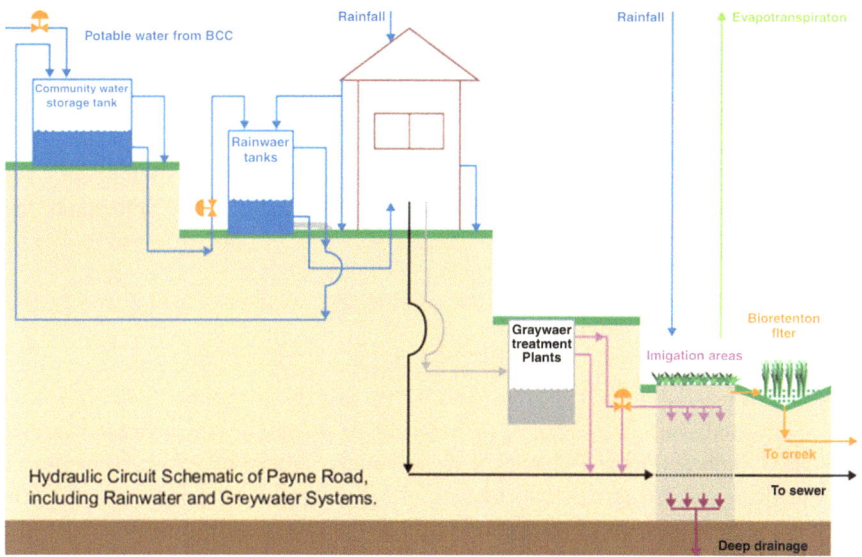

Fig. 10 Water schematic for Payne Road (*Source* Davis and Farrelly 2009)

Description of the project: The Payne Road development in Brisbane's suburban fringe demonstrates an integrated approach to water services that include rainwater harvesting, greywater recycling and demand management. The decentralised systems are integrated with the centralised system as rainwater tanks receive back-up from the main supply and blackwater is discharged off-peak to an adjacent sewer main (Davis and Farrelly 2009) (Fig. 10).

4.1 What Are the Main Features?

Rainwater tanks: Each household is equipped with an 18–22 m^3 rainwater tank for each household. There are two 75 m^3 communal tanks, located at the bottom of the subdivision, designed for storage of household rainfall excess, provision for fire-fighting and future supply of households. The communal tanks were.

Greywater treatment: Greywater is treated by an aerobic composting system. The systems are installed on individual houses and they convey treated water to gardens via subsurface irrigation. Moisture sensors detect saturation levels and excess treated greywater is discharged into the sewer (Gardner et al. 2006). Wastewater from kitchens and toilets is conveyed via low-infiltration, reticulated gravity sewerage to a communal sewer pump well. As the sewer main already operates at peak flow capacity, sewage from the development is withheld in a pump well until it can be discharged during off-peak times into the main sewerage trunk (Gardner et al. 2006).

Bioretention basins: The stormwater run-off not captured by rainwater tanks is captured in swales and treated in a bioretention system at the bottom of the development.

Stormwater sensitive landscape: The landscape of the development ensures that stormwater drains to the bottom of the subdivision where a bioretention basin allows controlled discharge off the site and eventually back into Enoggera Creek. The basin is 25 m wide and 80 m long with a central trench filled with sandy loam and incorporates four crushed-rock filters and two 0.5 m high rock weirs.

4.2 What Were the Strengths?

Drought conditions in southeast Queensland have shown that decentralised subdivisions such as Payne Rd, with a water supply system based on rainwater collection are a viable alternative to municipal water supplies. Between May 2006 and December 2007, rainwater contributed almost 80 % of all water consumed at Payne Road in spite of the reduction in average annual rainfall by 40 %.

Wastewater at Payne Road allowed approximately 75 % of the greywater generated to be reused for irrigation purposes. Moreover, rainwater quality after treatment meets all of the current Australian Drinking Water standards. Feasibility studies demonstrated that increased costs of implementing unconventional technologies were lower than projected rates for extending Brisbane City Council water and sewerage infrastructure to the development.

4.3 What Were the Challenges?

Rainwater pumping and treatment systems within the houses are relatively inefficient compared to business as usual and manufactured water sources. The specific energy at Payne Road of around 5 kWh/kL is 10 times higher than business as usual main supply at 0.5 kWh/kL. However, when looking at the overall household energy use, the rainwater systems use 7.5 % of the total household energy use. Moreover, the direct energy use of the household is only around 20 % of the greenhouse budget attributable to the house, suggesting that considerably greater energy savings are available to householders by changes in other urban practices. For example, producing or procuring local grown foods or using public transport will rapidly replace the deficit of an inefficient rainwater pumping system. Other suggestions on how to offset these energy losses include greenhouse gas efficient hot water system and grid connected photovoltaics (PVs).

Over instrumentation in the home and on-site: The many different monitoring systems have generated excessive amounts of data, which have proved unnecessary. Furthermore, over instrumentation was also linked to the issue of increased costs required to maintain monitoring and evaluation systems. Increased cost relative to

Fig. 11 Aerial view of BedZED (*Source* BedZED 2007)

time delays in the approvals process as well as with the implementation of unconventional technologies.

5 BedZED: Zero Energy Development

Beddington Zero Energy Development (BedZED) is an environmentally friendly housing development near Wallington, England, in the London Borough of Sutton. It was designed by the architect Bill Dunster, who was looking for a more sustainable way of building housing in urban areas (BedZED 2007; Fig. 11).

5.1 What Was the Aim of the Project?

- Reduce the overall consumption of potable water with the installation of water efficient appliances.
- Raise awareness and make occupants take responsibility for their own consumption and be able to monitor it.
- Installation of rainwater harvesting system by draining surplus water from the slightly arched green roofs.
- Wastewater treatment.

5.2 What Are the Main Features?

- **Zero energy**—The project is designed to use only energy from renewable sources generated on-site.
- **Energy efficient**—The houses face south to take advantage of solar gain, are triple-glazed and have high thermal insulation.
- **Water efficient**—Most rainwater falling on the site is collected and reused. Appliances are chosen to be water efficient and use recycled water when possible. A 'Living Machine' system of recycling wastewater was installed, but is not operating. A 'Living Machine' (LM) (as designed by Living Technologies (LT)) in a greenhouse located in the BedZED services building for the purpose of full on-site wastewater treatment would:

  (i)   supply treated effluent for landscape irrigation and the sports field,
  (ii) supply treated effluent for reuse in the toilets in the clubhouse,
  (iii) act as a small botanical nursery and an educational resource for the site and residents,
  (iv) grow plants in the LM for production of essential oils.

- **Low-impact materials**—Building materials were selected from renewable or recycled sources within 35 miles of the site, to minimise the energy required for transportation.

5.3 What Were the Strengths?

- Green roof concept (stormwater attenuation, rainwater harvesting)
- Water conservation measures on the potable supply
- Sky gardens (subsurface trickle irrigation)

5.4 What Were the Main Challenges and Results? (Shirley-Smith and Butler 2008)

- Costs higher than anticipated (labour, energy, extra equipment and water quality testing).
- Values of properties reported to be 15 % higher than similar adjacent properties.
- At the local (design) level, the concepts of demand management, rainwater harvesting, stormwater management, green water recycling and thermal heat storage, which were supposed to be combined within the same system, were never really integrated into the same design concept. This led to oversized underground tanks, wasted treated effluent, large amounts of main water top-up and uncertainty about the quality of water supplied for various purposes.
- At the organisation level, the project was handled by different parties, which was inefficient and error-prone.

5.5 *Which Are the Key Lessons?*

- The need to place a single organisation that will take responsibility for all aspects of integrated water management.

6 Multiple-Use Water Services: Winrock's Experience in Africa and Asia

In many districts around the world, people living in poverty have limited or no access to substantial water services in order to cover a variety of vital needs such as drinking, hygiene, cooking, irrigation, livestock production and small-enterprise maintenance.

6.1 *Multiple-Use Water Services (MUS)*

MUS is an integrated, participatory service delivery approach that considers water users' multiple needs as a starting point and involves design, financing, provision and management of sustainable water services along with supporting programmes regarding human health, livelihoods and environment. MUS approaches can potentially benefit more than 1 billion people in rural South Asia and sub-Saharan Africa (Renwick et al. 2007) by increasing income and reducing poverty, improving human health, livelihoods and social equity.

Winrock International is a non-profit organisation with considerable experience and involvement in Multiple-Use Water Services through various programmes and partnerships (WI 2014). Winrock targets sustainable strategies within communities, watersheds and regions in many countries around the world to assist less advantaged people, reinforce economic activities and conserve natural resources.

In Table 1, an overview of the Winrock International experience in multiple-use water services is presented, followed by a short description of the interventions implemented in selected cases.

6.2 *Nepal*

The implementation of MUS approach in Nepal occurred mainly through the Smallholder Irrigation and Market Initiative (SIMI) and the Education for Income Generation (EIG), projects primarily funded by US Agency for International Development (USAID), with main partners Winrock International and International Development Enterprises (IDE) as well as other local partners including the Center

Table 1 Overview of Winrock's MUS programmes in seven countries

Country /programme	Clients/beneficiaries	Target		
		Household	Community	Catchment
Nepal	12,500		x	x[a]
India	7,250		x	
Tanzania	68,000	x	x	
Rwanda	80,000	x	x	x
Niger—WAWI	13,500		x	
Niger—WA-WASH	10,000	x	x	
Burkina Faso	22,000	x	x	x?

Source Renwick (2012)

[a] Estimated based on actual and targets (for ongoing programmes). Self-supply is excluded

for Environmental and Agricultural Policy Research, Extension and Development (CEAPRED), Support Activities for the Rural Poor (SAPPROS) and the Agricultural Enterprise Center (AEC).

MUS constructed in the Middle Hills of Nepal consisted of piped gravity flow systems that provided sufficient and clean water for domestic uses and crop growing in households, improved with drip irrigation systems (SIMI 2009). The MUS systems incorporated intakes, reservoir tanks as well as transmission and distribution networks. As the systems built were small and used water from small spring sources, soil erosion, landslides and other ecological problems were not encountered. Before the construction orientation sessions, feasibility studies and field surveys took place.

Multiple-use water services linked with micro-irrigation systems constructed in Nepal proved to be not only successful in meeting both domestic and irrigation demand, but also cost-effective investments, providing an annual income gain of $200 per household coming from a $100 initial investment (SIMI 2009; Fig. 12).

6.3 Tanzania

Multiple-use water services have been introduced in Tanzania by the Integrated Water, Sanitation and Hygiene (iWASH) programme in 2010. iWASH is a USAID funded project, implemented by Winrock International, Florida International University, CARE Tanzania and WaterAid Tanzania, to assist people in Tanzania in improved access to clean water, hygiene and sanitation services (GLOWS 2012).

iWASH incorporates the MUS approach to provide sustainable water supply to rural and small town inhabitants based on actual household and community water needs. Under this approach, low-cost technologies for multiple uses were developed, such as locally manufactured rope pumps, rota-sludge drilling and water filters. Encouraging results showed that through iWASH more than 51,000 people

Fig. 12 MUS in Nepal (Photo courtesy: USAID and Winrock International)

in specific areas in Tanzania gained access to drinking water and nearly 11,500 people benefit from better sanitation services. The programme also involves hygiene promotion projects, trainings and capacity building workshops.

6.4 The WA-WASH Project

West Africa Water, Sanitation and Hygiene Initiative (WA-WASH), is a USAID-supported programme that aims to improve water supply, hygiene and sanitation services for low-income populations in Burkina Faso, Niger, Ghana and Mali (WI 2014).

Following the MUS approach, the provision of water services is designed according to the domestic and productive water needs of the dwellers. Gender behaviour towards water uses and household investment in water services is considered in order to achieve improved social equity. Water infrastructure is combined with supporting health, livelihood and environmental activities benefiting around 34,000 people. Such activities include hygiene and sanitation promotion, education and training to local farmers (e.g. on soil and water conservation techniques) or to private sector small businesses, such as well-drilling teams, pump fabricators and installers (Fig. 13).

Fig. 13 A well-drilling team in Burkina Faso (*Source* WI 2014)

6.5 What Are the Lessons Learned from Winrock's Experience in MUS?

- Significant potential for impact in increasing income, mitigating poverty, and improving human health, livelihoods and social amenity.
- Improved benefits by linking MUS with supporting health, livelihood and environmental programmes and capacity building activities.
- Significance of regulatory frameworks and financing structures to support the MUS approach.
- Cooperation between involving organisations, private sector agencies, local authorities, water suppliers and water users.

Keywords and Definitions

BedZED	Beddington Zero Energy Development is an environmentally friendly housing development near Wallington, England, in the London Borough of Sutton
DRMS	Decentralised Rainwater Management Systems
Green City, Clean Waters' plan	25-year plan adopted by the city of Philadelphia to reduce pollution from CSO to the extent that 85 % of the water would be treated in the CSS
Green Stormwater Infrastructure (GSI)	Applies nature-complying methods to manage run-off at the source through soil and vegetation absorption and filtration of water

Land–Water-Infrastructure	Implements the following practices large-scale application of GSI techniques on public areas; large-scale tree planting; increased access and enhanced recreational opportunities on waterways; open spaces maintenance and use to manage stormwater; redevelopment of empty and abandoned areas and transformation to open spaces; restoration of streams for improvement of aquatic habitats; grey stormwater infrastructure maintenance and improvement projects
MUS	Multiple-Use water Services
RHM	Rainwater harvesting and management
RWH	Rainwater harvesting
WSUD	Water sensitive urban design

References

BedZED (2007) Zero energy development. http://www.zigersnead.com/current/blog/post/bedzed-beddington-zero-energy-development/11-12-2007/351/. Accessed 20 December 2013

Brown R, Clarke J (2007) Transition to water sensitive urban design: the story of Melbourne. Australia. Report no. 07/01. Facility for Advancing Water Biofiltration, Monash University, Clayton

CDM Smith (2013) Official website. http://cdmsmith.com/. Accessed April 2014

Davis C, Farrelly M (2009) Demonstration projects: case studies from South East Queensland, Australia. National Urban Water Governance Programme, Monash University, Australia. http://www.urbanwatergovernance.com/pdf/demo_proj_se_qld.pdf. Accessed 2013

Farrelly MA, Davis C (2009) Demonstration projects: case studies from Melbourne, Australia. National Urban Water Governance Programme, Monash University, Clayton

Gardner T, Millar G, Christiansen C, Vieritz A, Chapman H (2006) Urban metabolism of an ecosensitive subdivision in Brisbane, Australia. In: Proceedings of Enviro 06, May 9–11, 2006, Melbourne

GLOWS (2012) Tanzania Integrated Water, Sanitation and Hygiene Programme (iWASH). Global Water for Sustainability Programme. http://www.globalwaters.net/wp-content/uploads/2012/02/August-21-NEW-iWASH-brief.pdf. Accessed December 2013

Han M (2008a) Rainwater harvesting system at Star City as an example of eco-efficient and climate change adaptation. In: 1st regional workshop of eco-efficient water infrastructure in Asia, November 10–11, 2008, SETEC, Seoul, Korea

Han M (2008b) Rainwater management and rainwater use examples in Seoul. Seoul, Korea. http://www.kankyo.metro.tokyo.jp/en/attachement/seoul.pdf. Accessed December 2013

Han M (2012) Smart water portfolio & rainwater revolution: from Drain City to Rain City—a case study of Korea. European Union conference, Vienna, Austria

Han M, Park J (2005) Rainwater water management in Korea: public involvement and policy development. In: International workshop on rainwater and reclaimed water for urban sustainable water use, Tokyo, Japan

Han MY, Mun JS (2011) Operational data of the Star City rainwater harvesting system and its role as a climate change adaptation and a social influence. Water Sci Technol 63(12):2796–2801. IWA Publishing

International Water Association (IWA) (2008) Seoul's Star City: a rainwater harvesting benchmark for Korea. Water 21:17–18

Lloyd SD (2001) Water sensitive urban design in the Australian context. Technical report 01/7, Melbourne water, Melbourne

Lloyd SD (2004) Exploring the opportunities and barriers to sustainable stormwater management practices in residential catchments. Cooperative Research Centre for Catchment Hydrology, Melbourne

PWD (2013) Philadelphia Water Department. Official website. http://www.phillywatersheds.org/. Accessed April 2014

Renwick M (2012) What does MUS look like? Moving from theory to practice in 7 countries. Presentation from the 2012 World Water Week in Stockholm. Winrock International. Stockholm Water Week, August 30, 2012. http://www.worldwaterweek.org/documents/WWW_PDF/2012/Thur/Scaling-pathways-for-multipleuse/Renwick.pdf. Accessed December 2013

Renwick M et al (2007) Multiple use water services for the poor: assessing the state of knowledge. Winrock International, Arlington, VA. http://www.winrockwater.org/docs/Final%20Report%20Multiple%20Use%20Water%20Services%20Final%20report%20feb%2008.pdf. Accessed December 2013

Shirley-Smith C, Butler D (2008) Water management at BedZED: some lessons. Proc Inst Civil Eng Energy 161(2):113–122. doi:10.1680/ensu.2008.161.2.113

SIMI (2009) Nepal smallholder irrigation market initiative (SIMI) project completion report. June 1, 2003–September 30, 2009. USAID Cooperative Agreement no. 367-A-00-03-00116-00. http://pdf.usaid.gov/pdf_docs/PDACP099.pdf. Accessed December 2013

UN ESCAP (2012). Low carbon green growth roadmap for Asia and the Pacific, background policy paper. Water resource management: policy recommendations for the development of eco-efficient infrastructure. Prepared by: Dr Reeho Kim, Dr Jungsoo Mun and Dr Jongbin Park, Korea Institute of Construction Technology. United Nations publication 2012

UNEP & SEI (2009) Rainwater harvesting: a lifeline for human well-being. United Nations Environment Programme and Stockholm Environment Institute. http://www.unwater.org/downloads/Rainwater_Harvesting_090310b.pdf. Accessed December 2013

WI (2014) Winrock International, official website. http://www.winrock.org. Accessed 28 April 2014

Wong THF (2001) A changing paradigm in Australian urban stormwater management. Keynote address: 2nd South Pacific Stormwater Conference, Auckland, New Zealand, 2001, pp 1–8